W9-CBK-534

Electronic Components Handbook

Electronic Components Handbook

Thomas H. Jones

RESTON PUBLISHING COMPANY, INC.

A Prentice-Hall Company
Reston, Virginia

Library of Congress Cataloging in Publication Data
Jones, Thomas H
    Electronic components handbook.

    Includes bibliographies and index.
    1. Electronic apparatus and appliances –
Handbooks, manuals, etc.   I.   Title.
TK7870.J65   621.3815   77-22341
ISBN 0-87909-222-X

© 1978 by
Reston Publishing Company, Inc.
A Prentice-Hall Company
Reston, Virginia 22090

All rights reserved. No part of this book may be reproduced in any way,
or by any means, without permission in writing from the publisher.

10   9   8   7   6   5   4   3   2

PRINTED IN THE UNITED STATES OF AMERICA

Contents

List of Tables

Preface

Components, or parts, whichever you prefer to call them (in this handbook, the terms are used interchangeably) — resistors, capacitors, relays, transformers, semiconductors, and all the other discrete devices are the building blocks of electronic equipment of all types. This handbook has been written to provide component users — designers, engineers and technicians — a better understanding of component performance characteristics so that they will be able to select the best component to meet their own specific requirements. Equations, data and theory of interest only to component designers have been kept to a bare minimum.

There are thousands and thousands of electronic components available; it would be impossible to describe them all. Preference has been given to industrial varieties over strictly Mil-spec. parts on the one hand, and over low-cost hobbiest parts at the other extreme; and preference has also been given to parts that are readily available as stock items, and particularly to those available through mail-order catalogs. Within this framework, parts are divided into three quality levels; military, those parts that essentially meet Mil-specs.; industrial, those parts used in communications, industrial control, automation, and data processing equipment; and commercial, those parts used in consumer products.

Component selection is always a compromise between performance, reliability and cost. All three factors are important. Including costs, even approximations, in a handbook is risky in today's economy.

Costs as given in this handbook are for small quantities, and are for comparative purposes for aid in making a selection from a group of similar components.

The bibliography at the end of each chapter contains sources of additional user-oriented component information. Many of the catalogs are handbooks in themselves.

Acknowledgments

Writing a handbook like this is not a solo venture. I have received the generous help of many components manufacturers who provided me with catalogs, engineering and technical bulletins, handbooks and other published material, with permission to use some of their published material, and with illustrations. I thank all of these companies, listed below, for their invaluable help.

Allen-Bradley
Amphenol Component Group,
    Bunker Ramo
Arco Electronics, Inc.
Aerovox Industries, Inc.
A. W. Haydon, Division of North
    American Phillips Co.
The Carborundum Company
Centralab Electronics Division,
    Globe-Union, Inc.
Clarostat Mfg. Co., Inc.
Cornell-Dubilier Electronics
    Division, Federal Pacific
    Electric Company
Corning Glass Works
Cramer Controls, Conrac Corp.
Dale Electronics, Inc.
Elco Corp.
Erie Technological Products, Inc.
Fenwal Electronics

General Electric, Electronic
    Components Sales
General Instrument Corp.,
    Capacitor Division
GTE Automatic Electric
ITT Cannon Electric
James Millen Mfg. Co., Inc.
JFD Electronics Corp.
J. W. Miller Coil Co.
LFE Corporation
LICON Division, Illinois Tool
    Works, Inc.
Magnecraft Electric Co.
Mallory Capacitor Co., Division of
    P. R. Mallory & Co., Inc.
Matrix Science Corporation
Micro Switch Division, Honeywell
Microtran Co., Inc.
Motorola Semiconductor
    Products, Inc.

N L Industries, Inc.
Nytronics, Inc.
Oak Industries, Inc., Smith Division
Ohmite Mfg. Co., North American
    Phillips Co.
Pico Electronics, Inc.
Potter & Brumfield, Division of
    AMF, Inc.
RCA
Robertshaw Controls Co.
Sangamo Electric Co., Capacitor
    Division
Sigma Instruments, Inc.
Spectrol Electronics Corp.

Sprague Electric Co.
Stackpole Components Co.
Struthers-Dunn, Inc.
Switchcraft, Inc.
Teledyne Relays
Tempo Instrument, Inc.
Texas Instruments, Inc.
Thermometrics, Inc.
Thirdarson & Meissner, Inc.
TRW Electronic Components
    Division
Victory Engineering Corp.
YSI-Sostman

   Special thanks are due to the General Electric Company Space and RESD Libraries at the Valley Forge Space Center where much of the research that went into the handbook was done, and to Eugenia Sowicz, Specialist/Technical Information on the library staff whose assistance made the work a lot easier.

Electronic Components
Handbook

CHAPTER ONE

Fixed Capacitors

Capacitors are electronic components which have the ability of storing electrical energy. Basically, all capacitors consist of two parallel, facing conductive surfaces separated by an insulating material called the dielectric. In the simplest form of capacitor, two equal-sized metal plates are separated by an air gap. Fixed capacitors (Fig. 1–1) are discussed in this chapter; variable capacitors in Chapter 2.

The capacitance of a capacitor varies directly with the area of the facing plates and the dielectric constant of the insulating material. Capacitance also varies inversely with the separation of the plates, in other words, with the thickness of the dielectric (Fig. 1–2).

Many materials are used as capacitor dielectric materials and each has its own dielectric constant. Examples of dielectrics include air, glass, mica, ceramics, plastics, paper, oil, oxides of certain metals and the non-material vacuum (Table 1–1).

A direct current cannot flow through the dielectric material in a capacitor. Alternating current cannot flow through a capacitor either, but the electrons which comprise the alternating current flow move to and from the individual capacitor plates and *simulate* an apparent flow of current through the capacitor. An ac meter indicates ac flow to and from the capacitor and gives the appearance of a continuous current loop as with dc. Hence, a blocking or a coupling capacitor can separate ac and dc components of a signal in a circuit.

The amount of alternating current "flowing" through a capacitor increases with frequency; a capacitor can be used as a low- or high-pass filter to distinguish between two signals of different frequencies. Combined with an inductance, a capacitor will form a

Fig. 1–1 Common types of fixed capacitors. *(Courtesy of Sprague Electric Company)*

tuned circuit to null or peak a signal of some given frequency. With a resistor to limit the rate at which current flows into a capacitor, a timing circuit is formed. You can charge a capacitor slowly and discharge it almost instantly to power an electronic flash unit, or you can use a capacitor in a dc power supply to filter, or smooth out, a pulsating direct current.

Fixed capacitors used in electronic equipment can be grouped into eight basic types by the dielectric material used. (Variable capacitors are covered in Chapter 2.)

Paper Capacitors. Paper has been used as a capacitor dielectric for a long time, and it is still used extensively for high-voltage and high-discharge current applications where the advantage of a thick porous dielectric that can retain a liquid to suppress corona is more important than small size. For low-voltage capacitors (600 vdc and under) paper has been almost completely displaced by plastic films and film-paper combinations.

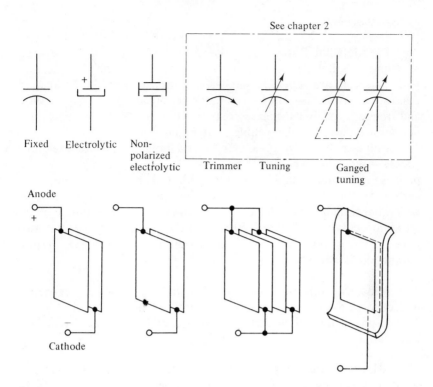

Fixed Electrolytic Non-
polarized
electrolytic Trimmer Tuning Ganged
tuning

Anode
+

Cathode

Fig. 1–2 Capacitors. Symbols for **(a)** fixed capacitors, **(b)** variable capacitors. The capacitance of a simple air-dielectric capacitor **(c)** can be increased by positioning the plates closer together **(d)**, increasing the number of plates in parallel **(e)**, or by substituting a material with a higher dielectric constant such as one of the plastic-film dielectrics **(f)**.

TABLE 1-1   Dielectric Constants

Dielectric Material	Dielectric Constant	Dielectric Material	Dielectric Constant
Air	1.0001	Mica	4.5 to 7.5
Vacuum	1.0000	Glass	6.7
Paper (Impregnated)	4.6	Green Glass	8.3
Polyester	3.0	Quartz	4.2
Polystyrene	2.5 to 2.7	Steatite	5.5 to 6.5
Polycarbonate	3	Titanium Dioxide	80 to 120
Polypropylene	2	Barium Titanate	200 to 16,000
Polysulfone	3	Aluminum Oxide	10
Teflon	2	Tantalum Oxide	11
Polyimide	3.5		

Plastic Film Capacitors. A variety of plastic films are used as capacitor dielectrics. Compared to paper dielectric capacitors, film capacitors are generally smaller, and less susceptible to moisture infiltration problems. They may have better or worse temperature and frequency stability characteristics. All plastic films are not alike, neither are they interchangeable in all applications.

Polyesters (Mylar ®, Kodar ®) and polystyrene are the most widely used films—polyester in general-purpose applications, and polystyrene where stability is required. Other films include polycarbonate, which is generally better than polyester; polypropylene, which has a higher working voltage capability; and polysulfone, polytetrafluoroethylene (Teflon TFE ®), and polyimide. These last two have higher operating temperatures. Paper and films are also used in various combinations.

Mica Capacitors. Natural mica, the second-oldest capacitor dielectric material, has significant advantages: It is inert, it will not change physically or chemically with age, and it has good temperature stability.

Glass Capacitors. Glass was the first dielectric material used in a capacitor. The glass capacitor was invented in 1746 by Pieter van Musschenbrock. He called it a Leyden Jar.

Modern glass capacitors were developed during World War II to reduce dependence on foreign sources of high-quality mica, principally India. The construction of glass capacitors is similar to that of mica capacitors.

Ceramic. The dielectric of a ceramic capacitor is usually a steatite or barium titanate ceramic; other ceramics are also used. A wide range of temperature coefficients and dielectric constants are obtained by varying the ceramic's composition.

Ceramic capacitors are the most widely used of all types because of their low cost, small size, wide range of characteristics, excellent high-frequency performance, and inherent reliability.

Electrolytic Capacitors. These capacitors are internally different from other types of capacitors in that the dielectric is not a separate material but an aluminum oxide or a tantalum oxide layer a few millionths of an inch thick formed electrolytically on the metallic anode.

The anode may be a foil or a pellet of sintered metal powder. The cathode is a liquid or solid electrolyte. A metal "cathode" serves merely as a conductor to provide a connection to the external circuit. The outstanding advantage of electrolytic capacitors is the large capacitance that can be achieved in a small volume.

Oil-Filled Capacitors. Applications involving high currents at power-line frequencies, high to extremely high-voltage dc filtering, and high-energy discharge cycles use capacitors having in common an oil impregnant, relatively large size, steel cases, and a paper or combination paper and polyester dielectric.

Each of the three applications require a separate class of capacitor: Alternating-current oil, direct-current oil, and energy discharge. Although similar in appearance, the three types are not interchangeable.

Air-Dielectric Capacitors. Fixed value air-dielectric capacitors are principally used as laboratory standards for calibration and measurement purposes. With precision construction and appropriate materials, a capacitance stability of 0.01% can be obtained over periods of years.

Vacuum and Gas-Filled Capacitors. Fixed vacuum capacitors are used as high-voltage capacitors in airborne transmitting equipment or as blocking and coupling capacitors in high-voltage industrial and communications equipment. Since the dielectric constant of vacuum is essentially the same as for air, the capacitance-to-volume ratio is rather low. Gas-filled types are used for very high voltages — on the order of 250,000 volts.

Frequency Characteristics. All capacitors are limited in their range of operating frequencies. An ideal capacitor would have pure capacitive reactance, but in the real world, imperfect dielectrics, case materials, and internal, package, and lead configurations rule otherwise.

A capacitor, as a practical device, has not only capacitance, but also resistance and inductance (see Fig. 1–3). These factors must be carefully considered in selecting capacitors.

Lead inductance and the resistance of the leads and plates put a limit on high-frequency performance. As frequency increases, the voltage drop across R_1 increases, as does the lead inductive reactance. At the resonant frequency $(X_L = X_C)$, the impedance of the capacitor is $R_1 + R_2$. Above the resonant frequency, the capacitor behaves as an inductor. There is no low-frequency limit on the performance of a capacitor, only an inappropriateness of its capacitance

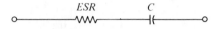

(a) Capacitor equivalent circuit

(b) Equivalent circuit for
electrolytic capacitors

C = Capacitance
L = Inductance
R_S = Series resistance (losses in capacitor)
R_P = Parallel resistance (insulation resistance)
ESR = Equivalent series resistance
        (combines R_S and R_P)

Fig. 1–3 Simplified capacitor equivalent circuits; **(a)** all capacitors, **(b)** electrolytic capacitors.

value to the circuit and an impracticability to manufacture that particular style or type of capacitor in a usable value. (See Fig. 1–4.)

Low insulation resistance limits the ability of a capacitor to hold a charge. This presents a problem in timing and coupling circuits, but not in most power filtering applications.

1.1 CAPACITOR DEFINITIONS

ANODE: The positive electrode of a capacitor.

CAPACITANCE: The property of a capacitor to store electrical energy when voltage is applied. It is measured in farads, microfarads, and picofarads. Farad is seldom used as it is so huge a quantity.

CAPACITANCE PURCHASE TOLERANCE: The part manufacturer's guaranteed maximum deviation (at a time of purchase) from the specified nominal value, at standard environmental conditions, given in percent.

CAPACITIVE REACTANCE: The opposition offered to the flow of an alternating or pulsating current by capacitor, measured in ohms.

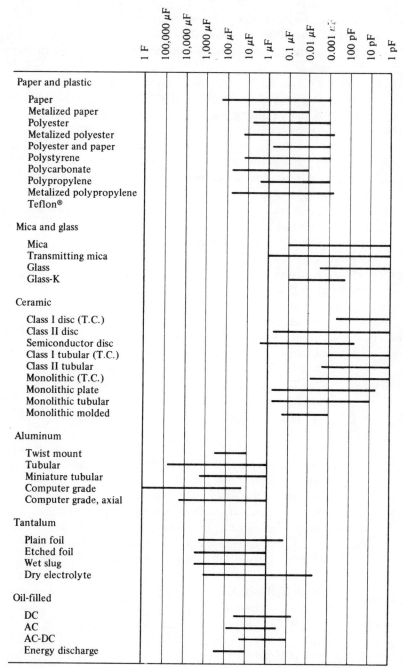

Fig. 1-4 Range of capacitances generally available in capacitors on the basis of dielectric material used.

CATHODE: The negative electrode of a capacitor.

CORONA: The ionization of air or other vapors which causes them to conduct current, a phenomena caused by high voltage or high-voltage gradients.

DIELECTRIC: The insulating material between the plates of the capacitors. Typical dielectrics include air, paper, mica, plastic, oil, and oxides of aluminum and tantalum.

DIELECTRIC ABSORPTION: The property of an imperfect dielectric whereby all electric charges within the body of the material caused by an electric field are not returned to the field. This produces a time lag in charging and discharging. Also called *dielectric hysteresis.*

DIELECTRIC CONSTANT: The characteristic of a dielectric material that determines how much electrostatic energy can be stored per unit volume when a voltage is applied, given as a ratio to a vacuum dielectric.

DISSIPATION FACTOR (DF): The ratio of resistance to reactance, measured in percent.

ELECTROLYTE: The current-conducting solution (liquid or solid) between two electrodes or plates of a capacitor.

EQUIVALENT SERIES RESISTANCE (ESR): The sum of all internal series resistances concentrated or "lumped" at one point in the capacitor equivalent circuit and treated as one resistance.

IMPEDANCE (Z): Total opposition offered to the flow of an alternating or pulsating current, measured in ohms. Impedance is the vector sum of resistance, capacitive reactance and inductive reactance.

IMPREGNANT: A substance, usually liquid, used to saturate a paper dielectric and to replace the air between the paper fibers. Impregnation increases both the dielectric strength and the dielectric constant of the assembled capacitor.

INSULATION RESISTANCE (IR): The direct current resistance measured across the capacitor terminals. For small paper, film, mica, and ceramic capacitors, IR is expressed in megohms and is usually in the order of 100,000 megohms. As capacitance value (and the area of dielectric) increases, the IR decreases proportionately. Insulation resistance is often specified in ohm-farads or, more commonly, megohm-microfarads.

LEAKAGE, DC (DCL): The direct current which flows through a dielectric capacitor when voltage is impressed across its terminals.

POWER FACTOR (PF): The ratio of resistance to impedance, measured in percent. For most practical purposes, the same as dissipation factor.

QUALITY FACTOR (Q): The ratio of reactance to resistance.

RIPPLE VOLTAGE (OR Current): The ac component of a uni-directional voltage or current, usually small in comparison with the dc component.

SCINTILLATION: Random transient variations in a capacitor.

SURGE VOLTAGE (OR Current): A transient variation in the voltage or current of large magnitude and short duration.

TEMPERATURE COEFFICIENT (TC): The change in capacitance of a capacitor per degree change in temperature. It may be positive, negative, or zero and is usually expressed in parts per million per degree Celsius (ppm/°C).

1.2 CAPACITOR MARKING

Most capacitors are plainly marked with purchase value, working voltage, and polarity or outside foil as appropriate. Some types are color-coded. In the past, color-code marking was more extensively used with more than one system of coding, although they all used the same color-digit code. Figures 1–5 and 1–6 show current marking for ceramic capacitors and molded mica capacitors.

1.3 PAPER CAPACITORS

Tubular paper capacitors (Fig. 1–7), and film capacitors are constructed in essentially the same manner, with variations depending on how the leads are to be attached to the foil. For paper capacitors, alternate strips of foil and paper are wound into a tight roll. Two or more layers of paper are used to guard against minute pinholes and conductive particles that cannot be eliminated completely in the paper manufacturing process. High-purity aluminum foil is used, after being carefully cleaned to eliminate any contaminants that could react with the impregnant and ultimately affect capacitor performance.

If the leads are to be attached to the foils by one or more tabs inserted in the roll, the paper is wider than the foil. In another method, foils and paper are wound offset so one foil extends out one end of the roll and the other foil out the other end. After winding, the extruded foils are crushed over the paper, and the leads are then either soldered or welded to the crushed foil ends. This method of construction effectively shorts the capacitor's inductance without affecting the capacitance, reducing the inductance of the capacitor almost to zero.

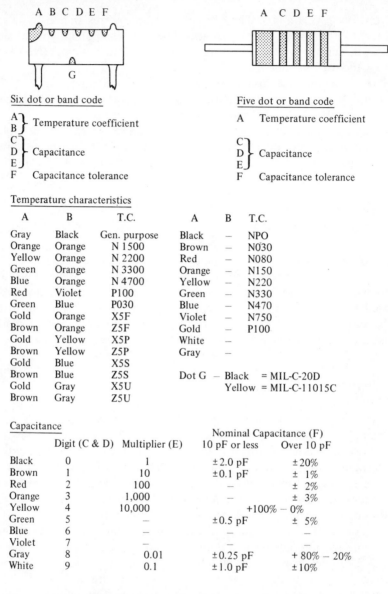

A B C D E F

G

Six dot or band code

A
B } Temperature coefficient

C
D } Capacitance
E

F   Capacitance tolerance

A C D E F

Five dot or band code

A   Temperature coefficient

C
D } Capacitance
E

F   Capacitance tolerance

Temperature characteristics

A	B	T.C.	A	B	T.C.
Gray	Black	Gen. purpose	Black	—	NPO
Orange	Orange	N 1500	Brown	—	N030
Yellow	Orange	N 2200	Red	—	N080
Green	Orange	N 3300	Orange	—	N150
Blue	Orange	N 4700	Yellow	—	N220
Red	Violet	P100	Green	—	N330
Green	Blue	P030	Blue	—	N470
Gold	Orange	X5F	Violet	—	N750
Brown	Orange	Z5F	Gold	—	P100
Gold	Yellow	X5P	White	—	
Brown	Yellow	Z5P	Gray	—	
Gold	Blue	X5S			
Brown	Blue	Z5S	Dot G	— Black	= MIL-C-20D
Gold	Gray	X5U		Yellow	= MIL-C-11015C
Brown	Gray	Z5U			

Capacitance

			Nominal Capacitance (F)	
	Digit (C & D)	Multiplier (E)	10 pF or less	Over 10 pF
Black	0	1	±2.0 pF	±20%
Brown	1	10	±0.1 pF	± 1%
Red	2	100	—	± 2%
Orange	3	1,000	—	± 3%
Yellow	4	10,000	+100% − 0%	
Green	5	—	±0.5 pF	± 5%
Blue	6	—	—	—
Violet	7	—	—	—
Gray	8	0.01	±0.25 pF	+ 80% − 20%
White	9	0.1	±1.0 pF	±10%

Note 1.   Nominal capacitance code is EIA-RS198. MIL-SPEC codes are not the same.

Note 2.   Five and six digit codes are both used for radial-lead and axial lead capacitors.

Note 3.   Disc capacitors normally have typographical marking but may be color coded.

Fig. 1–5   Tubular ceramic capacitor marking code.

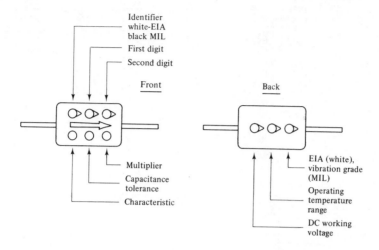

Color	Significant digits	Multiplier	Capacitance tolerance*	Characteristic	DC working voltage	Operating temperature	EIA/ Vibration
Black	0	1	±20%	–	–	−55°C to +70°C	10-55 Hz
Brown	1	10	± 1%	B	100	–	–
Red	2	100	± 2%	C	–	−55°C to +85°C	–
Orange	3	1,000	–	D	300	–	–
Yellow	4	10,000	–	E	–	−55°C to +125°C	10-2000 Hz
Green	5	–	± 5%	F	500	–	–
Blue	6	–	–	–	–	−55°C to +150°C	–
Violet	7	–	–	–	–	–	–
Gray	8	–	–	–	–	–	–
White	9	–	–	–	–	–	EIA
Gold	–	–	±0.5%**	–	1000	–	–
Silver	–	–	±10%	–	–	–	–

** Or ±0.5 pF, whichever is greater.
* ±1.0 pF

Fig. 1–6 Molded mica capacitor marking code.

After the leads are attached by either method, the roll is vacuum impregnated with a wax, plastic resin or a synthetic or mineral oil and then encapsulated to prevent moisture infiltration.

As a dielectric material, paper has its problems. Paper is extremely hygroscopic; despite plastic enclosures, housings and impregnants, moisture will be absorbed from the atmosphere, increasing power factor, decreasing insulation resistance and dielectric strength, and shortening life. For some critical applications, paper capacitors are supplied in hermetically sealed drawn-steel *bathtub* cases as shown in Fig. 1–8.

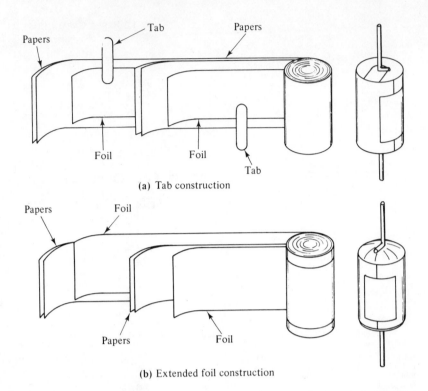

Fig. 1-7 Tubular paper capacitor construction. Film capacitors are constructed in essentially the same manner, **(a)** tab construction, **(b)** extended-foil construction.

Fig. 1-8 Paper capacitors in hermetically-sealed drawn steel bathtub cases. *(Courtesy of Nytronics Inc.)*

Metalized-Paper Capacitors.   It is possible to eliminate separate pieces of paper and foil in winding a capacitor by vapor-depositing a metal, such as zinc or aluminum on the surface of the paper.

Advantages include higher capacitance obtained from the same paper thickness because of closer contact between the electrodes and the paper, smaller overall size because of the thinner electrodes, and a higher voltage rating. Instead of an oil impregnant, a mineral wax or a rigid resin is used. This higher voltage rating comes about because of the self-healing characteristic of metalized capacitors. An internal arc through the paper dielectric burns away the extremely thin metal film, effectively clearing the short circuit.

The disadvantage of metalized paper capacitors is that they can be used only for low-current applications because of the limited current-carrying ability of the extremely thin film. This limitation includes all use where appreciable ac signals are present.

1.4 FILM CAPACITORS

Plastic films have been developed that are extremely resistant to moisture absorption, and can be manufactured in very thin gauges without pinholes or conducting particles. However, this is not a blanket statement that can be applied to all plastics used. (See Table 1–2.) Film capacitors are packaged in a variety of ways; epoxy-dipped, epoxy-molded, wrapped (an extension of the dielectric film is wrapped around the outside to form the case), and hermetically sealed in metal enclosures with glass-to-metal seals. (See Fig. 1–9.)

Polyester.   (Polyethylene pterapthalate) This plastic has achieved the most widespread use among plastic films as a capacitor dielectric. Polyester can be fabricated into very thin films; the combination of lower working voltage requirements and polyester's higher dielectric constant results in film capacitors much smaller in size than equivalent paper capacitors. Polyester film capacitors have almost completely replaced paper for most dc electronic applications, with operating temperatures as high as 125°C and working voltages up to 1600 vdc.

However, polyester capacitors do not have the temperature stability of paper—a 20% variation compared with 8%. But water absorption is only 0.8% compared with paper's 15%. These figures are approximates.

Construction of polyester film capacitors is essentially the same as for paper capacitors, except that no impregnant is used. It is possible to take advantage of the best qualities of both paper and poly-

TABLE 1-2 Film Dielectric Characteristics

Film Material	Polyester	Polystyrene	Polycarbonate	Polypropylene	Polysulfone	Polytetrafluor-ethylene	Polyimide	Paper (Ref)
Trade Names	Mylar, Kodar					Teflon	Kapton	
Temperature Range								
Low	−55°C	−55°C	−55°C	−55°C	−55°C	−55°C	−55°C	−55°C
High	+125°C	+150°C	+125°C	+105°C	+150°C	+170°C	+200°C	+125°C
Dielectric Constant								
60Hz	3.3	2.6	2.99	2.0	3.07	2.0	3.52	—
1kHz	3.2	2.6	2.99	2.0	3.07	2.0	3.52	—
1MHz	3.0	2.6	2.93	2.0	3.03	2.0	3.50	—
1kMHz	2.8	2.6	2.89	2.0	3.00	2.0	—	—
Dissipation Factor								
60Hz	.003	.001	.001	—	—	.001	—	.002 to .005
1kHz	.005	.0005	.0015	.0003	.0008	.0002	.003	.002 to .005
1MHz	.016	.0005	.010	.0003	.0034	.0002	.010	—
Insulation Resistance								
(MΩ-μF) 25°C	50,000	1,000,000	100,000	100,000	100,000	1,000,000	100,000	20,000
85°C	5000	100,000	10,000	10,000	10,000	—	—	200
125°C	1000	—	1000	—	1000	10,000	—	—
Dielectric Absorption	0.5 to 1%	0.02 to 0.05%	.5%	.1%	.2%	0.02% to 0.05%	—	0.6 to 5%
Water Absorption (24 hours)	0.8%	0.1%	0.3%	.005%	.22%	.01%	2.9%	15%

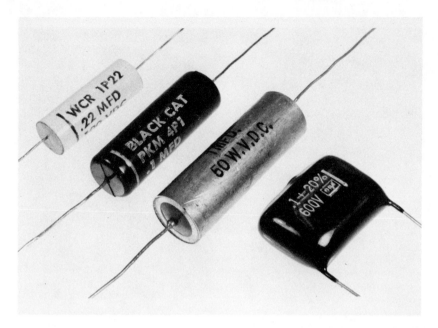

Fig. 1–9 Film capacitor case styles. Left to right: Wrapped, epoxy-molded, metal case with glass-to-metal seals, epoxy dipped. Bathtub cases are also used.

ester film by combining them in a dielectric. High dc working voltages can be obtained in relatively small size. See Fig. 1–10.

Polystyrene. This thermoplastic material was one of the first plastic films used in capacitors. The electrical characteristics of polystyrene are unsurpassed by any other plastic film now in general use. Polystyrene capacitors have excellent stability and can hold charges for extended periods of time. Some of this stability is due to the very low moisture pickup, 0.1%, and the absence of any chemical reaction between polystyrene with water. Polystyrene also possesses a uniform and slightly negative temperature coefficient, useful in matching positive temperature coefficients of other components.

Polystyrene must be used in comparatively thick films because of manufacturing difficulties with thin films. As the dielectric constant is also comparatively low, polystyrene capacitors run bigger than polyester capacitors. The top operating temperature is 85°C (polystyrene begins to melt at just over 90°C); for applications where temperature–capacitance retrace stability is required, the top usable temperature is restricted to 65°C.

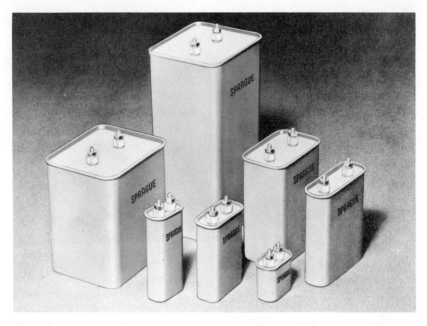

Fig. 1-10 High voltage paper/polyester film dielectric capacitors in drawn steel cases. Sprague type 271P shown have voltage ratings of 1000 to 12,500 vdc. These capacitors use a biodegradable synthetic polymer impregnant. *(Courtesy of Sprague Electric Co.)*

Polycarbonate. The insulation resistance, dissipation factor, and dielectric absorption of polycarbonate is superior to polyester, and the film can be made and used in thinner gauges. The change of capacitance with temperature is smaller, as the temperature coefficient approaches zero. The extremely thin gauge (as thin as 0.00008 inch) of polycarbonate film makes possible very small low-voltage capacitors. Moisture pickup is 0.3%, less than polyester. The low power factor of polycarbonate makes it ideal for ac applications.

Polypropylene. The outstanding characteristics of polypropylene film are its extremely low dissipation factor and a very high dielectric strength. Polypropylene also has a high insulation resistance and a more negative temperature coefficient than polystyrene. The moisture absorption is less than .01%, and there is no chemical reaction with water.

Polysulfone. Polysulfone is a dielectric material that has not quite arrived. It can withstand voltage at high temperatures better than most other plastics. Capacitors can be made with polysulfone film with temperature coefficients that approach zero. The dissipation factor is low and lessens with increasing temperature up to 150°C. Moisture absorption is 0.2%, which limits its stability; however, polysulfone does not react with water.

Polytetrafluoroethylene, PTFE–fluorocarbon (Teflon®). This plastic has a top operating temperature of +170°C; capacitors are stable and have extremely high insulation resistance. The disadvantages are high cost and large size.

Polyimide. Trademark named Kapton by DuPont and sometimes called H-film, polyimide capacitors have operating temperatures of up to 200°C. The film also has high insulation resistance, a low dissipation factor, and good stability. Its limitations are high cost and large capacitor sizes.

Metalized Film. Polyester and polycarbonate films can be metalized with zinc and aluminum to produce capacitors having both significantly smaller size and good reliability. They have the same self-healing characteristic as metalized paper and should also not be used in circuits where appreciable ac signals are present because of the limited current-carrying ability of the metalized plates.

The film used in the capacitor produces significant variation in capacitor physical size as shown in Fig. 1–11. This is in addition to

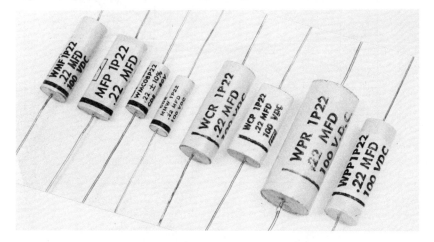

Fig. 1–11 Film wrap capacitors, variation in physical size. Left to right: Polyester, flat polyester, miniaturized polyester, metalized polyester, polycarbonate, flat polycarbonate, polystyrene, polypropylene. *(Courtesy of Cornell-Dubilier Electric Co.)*

the usual variation with capacitance value, working voltage, and foil or metalized construction. The capacitors also vary widely in range of available standard capacitance values and working voltages (see Table 1–3). There are also differences in capacitance versus temperature, insulation resistance versus temperature and dielectric strength versus temperature characteristics. See Figs. 1–12, 1–13, and 1–14.

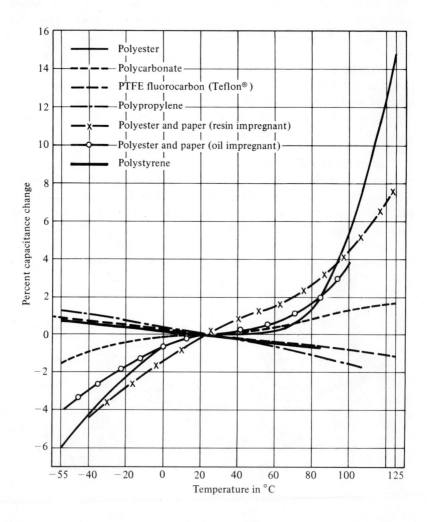

Fig. 1–12 Capacitance-temperature characteristics of plastic film capacitors.

TABLE 1-3   Summary Characteristics of Film and Paper Capacitors†

		Metallized Paper	Polyester	Metallized Polyester	Polyester and Paper	Polystyrene	Polycarbonate	Metallized Polycarbonate	Polypropylene
Case Styles:	Dipped		x	x	x	x			x
	Tubular	x		x	x	x		x	
	Wrapped		x	x	x	x		x	x
	Molded		x				x	x	
	Metal- & Glass-Sealed	x	x	x	x	x	x	x	x
	Bathtub	x		x		x			
Operating Temperature	Low	−55°C	−55°C	−55°C	−55°C	−55°C	−55°C	−55°C	−55°C
	High	+125°C	+85°C	+85°C	+85°C	+85°C	+125°C	+125°C	+85°C
	High, derated v.		+125°C	+125°C	+125°C	No	No	No	+125°C
Capacitance Tolerance	Standard	±20%	±20%	±20%	±10%	±10%	±10%	±10%	±10%
	Minimum	±5%	±1%	±1%	±2%	±1%	±1%	±1%	±0.25%
Capacitance	Minimum	.01 µF	.0001 µF	.001 µF	.001 µF	.0001 µF	.001 µF	.01 µF	.001 µF
	50	−	5.0	12.0	−	1.0	1.0	50.0	−
Maximum capacitance	100	5.0	5.0	18.0	−	1.0	1.0	22.0	2.0
at WVDC, dipped,	200	12.0	3.0	12.0	.5	.5	.5	12.0	.5
wrapped, and	400	2.0	1.5	2.5	.5	.5	.5	3.0	.5
tubular types	600	2.0	.22	1.5	.5	.22	.22	1.0	.22
	1000	−	.1	.1	.1	−	−	−	.05
	1600	−	−	.05	.05	−	−	−	.03
	2000	−	−	−	−	−	−	−	−
Stability (1 year)		±2%	±1%	±0.5%	±2%	±0.1%	±0.2%	±0.2%	±0.2%

†The table is limited to types of dielectrics generally available.

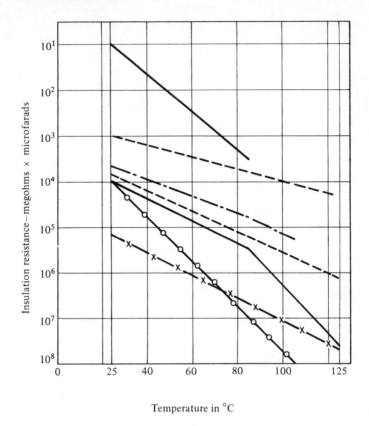

Temperature in °C

Fig. 1–13 Insulation resistance-temperature characteristics of plastic film capacitors.

1.5 MICA CAPACITORS

Mica is a mineral that occurs with granite and other igneous rock formations. Blocks of mica can be cleaved (split) into leaves as thin as one ten-thousandth of an inch. The sheets are uniform in thickness, the dielectric constant averages 6.85, and the dielectric strength is 3000 volts per mil.

Mica capacitors can be conveniently divided into two groups; small (Fig. 1–15), and transmitting (Fig. 1–16). Small mica capacitors are made two ways: a stacked plate construction in which sheets of mica and metal foil are interleaved, and a metalized construction in which silver electrodes are painted or screened and fired on the surface of the mica sheets. Transmitting micas are all made with the stacked plate construction.

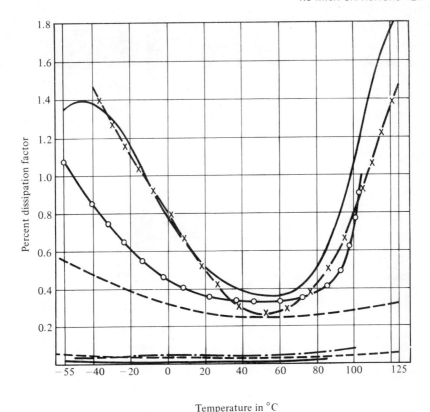

Fig. 1-14 Dissipation factor-temperature characteristics of plastic film capacitors.

Stacked-Plate Mica Capacitors. (See Fig. 1-17.) Mica sheets and foils (tin-lead, tin, brass, copper, or aluminum) are sandwiched with alternate foils extending beyond the ends of the mica sheets. The assembly is clamped, and lead wires are attached to the foils by spot-welding or soldering. The assembly is treated to prevent moisture absorption and then sealed in a molded or dipped case. A well-sealed case is important because mica capacitors are susceptible to moisture, which will increase capacitance and power factor and will lower insulation resistance.

Metalized Mica Capacitors, "Silver Micas." In this construction, silver electrodes are painted on sheets of thin mica in precise patterns, usually by silk-screening or spraying. The silver oxide and oil paint is fired, volatilizing the oil and reducing the oxide to

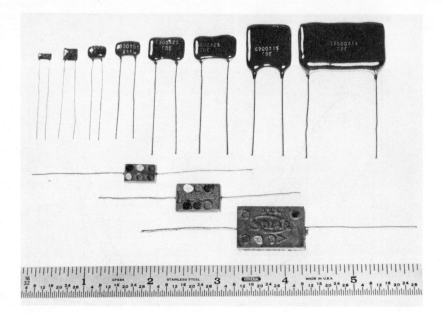

Fig. 1–15 Small mica capacitors. Above, dipped-epoxy cased; values range from 1pF to 91,000pF. *(Courtesy of Cornell-Dubilier Electric Co.)* Below, molded mica capacitors with color coded marking.

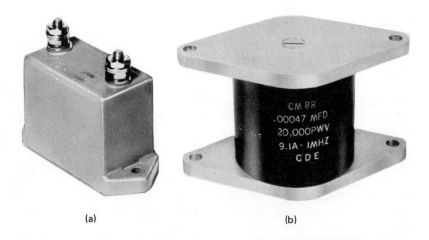

(a) (b)

Fig. 1–16 Transmitting mica capacitors **(a)** medium power, with voltage ratings up to 5,000 vdc, **(b)** high power, with voltage ratings 20,000 vdc. Capacitors are shown one third size. *(Courtesy of Cornell-Dubilier Electric Co.)*

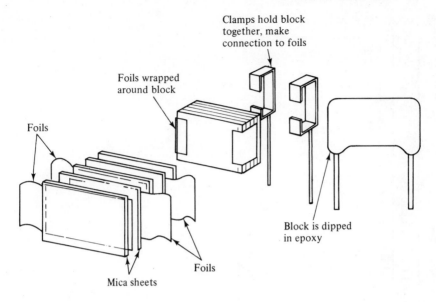

Fig. 1–17   Interleaved mica capacitor construction.

pure metallic silver. Sometimes the electrode painting is extended over opposite ends for lead attachment by soldering, but foils are generally inserted between the sheets to make better contact with the silver electrodes, and the lead wires are welded to these tabs.

Metalized micas are treated, usually with a silicone, to prevent moisture infiltration. They are then dipped in phenolic resin and epoxy-impregnated, rather than molded, as in the past.

The natural-material disadvantages of mica — variable uniformity and thickness, foreign sources — led to the development of synthetic mica. To make synthetic mica, a mixture of aluminum oxide, magnesium oxide, sand, and cryolite is heated to 1300°C and then slowly cooled over several days. Synthetic mica has cleavage planes similar to natural mica, but it is more difficult to split.

Natural and synthetic mica are used in several forms other than split leaves or sheets. Glass-bonded mica is made by combining finely powdered mica and a lead–borate glass binder. Small mica flakes and a silicon binder are combined to form reconstituted sheet mica, and synthetic mica can be pressed into a dense ceramic-like block.

Table 1–4 shows varieties of mica and glass capacitors.

TABLE 1-4 Typical Characteristics of Mica and Glass Capacitors

Style	Mica Dielectric						Glass Dielectric		
	Rectangular Molded, Axial Leads	Dipped, Radial Leads	Miniature Dipped, Radial Leads	Transmitting Plastic Potted	Transmitting Ceramic Cased	Transmitting Porcelain Cased	Rectangular, Sealed-in-Glass, Axial Leads	Rectangular, Sealed-in-Glass, Radial Leads	Glass-K (Comparable to Class II Ceramic) Tubular Axial Leads Sealed-in-Glass
Working Voltage	500 vdc	500 vdc 300 100	500 vdc 300 100	8000 vdc 6000 5000 4000 3000 2000 1500 1000 600 500 250	35,000 vdc to 1000 vdc	12,500 vdc 10,000 7000 5000 3500	500 vdc 300	300 vdc	50 vdc
Operating Temperature									
−55°C to +70 °C	x			x					
−55°C to +85 °C	x	x	x		x				
−55°C to +125°C	x	x	x			x	x	x	
−55°C to +150°C	x	x	x						x

TABLE 1-4   Continued

Style	Mica Dielectric						Glass Dielectric		
	Rectangular Molded, Axial Leads	Dipped Radial Leads	Miniature Dipped, Radial Leads	Transmitting Plastic Potted	Transmitting Ceramic Cased	Transmitting Porcelain Cased	Rectangular, Sealed-in-Glass Axial Leads	Rectangular, Sealed-in-Glass Radial Leads	Glass-K (Comparable to Class II Ceramic) Tubular Axial Leads Sealed-in-Glass
Temperature Characteristic and Capacitance Drift									
Not Specified	x	x	x						
Mica						x			
±200 ppm/°C	x	x	x	x	x				
±100 ppm/°C	x	x	x	x	x				
−20−+100 ppm/°C	x	x	x	x	x				
0−+70 ppm/°C	x	x	x	x	x				
+40−±25 ppm/°C							x	x	
+2−10%									x
+2−15%									x
+20−45%									x
Capacitance Min.	10 pF	1 pF	1 pF	47 pF	50 pF	50 pF	0.5 pF	1.0 pF	270 pF
Max.	0.02 μF	0.1 μF	470 μF	1.0 μF	1.0 μF	1.0 μF	0.01 μF	2400 pF	0.1 μF
Tolerance	±20%, ±10%, ±5%, ±2%, ±1%	±10%, ±5%, ±2%, ±1%, ±1%	±10%, ±5%, ±2%, ±1%	±5%, ±2%	±5%	±10%, ±5%	±20%, ±10%, ±5%, ±2%, ±1%	±20%, ±10%, ±5%, ±2%, ±1%	±20%, ±10%, ±5%

1.6 GLASS CAPACITORS

The rugged glass capacitor (Fig. 1–18) is stable and durable and is practically immune to temperature, aging, voltage, moisture, frequency shock, and vibration problems. Aluminum foil is used in place of the tin-lead foil. Glass is drawn into a one-mil thick flexible ribbon, layers of foil and glass are interleaved, leads are attached, and the assembly fused at a high temperature to form a monolithic structure of great physical strength. Stability and frequency characteristics are better than mica, but the cost is higher.

Fig. 1–18 Glass capacitors. Glass is used for both the dielectric and the fused case, providing a monolithic construction and true glass-to-metal seals. Corning type CY shown, values range from 0.5 pF to 0.01 μF. *(Courtesy of Corning Glass Works)*

1.7 CERAMIC CAPACITORS

Ceramic capacitors are manufactured in many shapes and sizes (Fig. 1–19) for a multitude of applications. The dielectric material of ceramic capacitors is a high-temperature, sintered, inorganic compound. Usually these materials are mixtures of complex titanate compounds such as barium titanate, calcium titanate, strontium titanate and lead niobate. Because of their wide variety of electrical characteristics, ceramic capacitors are divided into two classes.

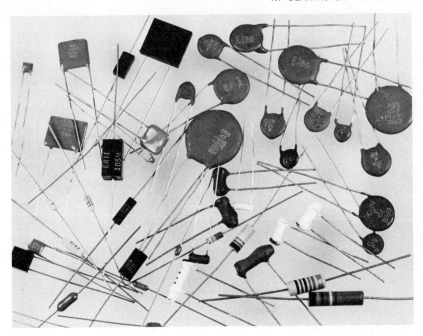

Fig. 1–19 Ceramic capacitors. *(Courtesy of Corning Glass Works and Erie Technological Products)*

Class I temperature-compensating ceramic capacitors have a predictable linear capacitance change with variations in operating temperature. This change is called *temperature coefficient of capacitance (TCC)* or *temperature coefficient (TC)* and is expressed in parts per million per degree Celsius (ppm/°C). This temperature coefficient is determined from a three-point measurement of capacitance at −55°C, +25°C, and +85°C or +125°C.

These Class I ceramic capacitors are identified by the nominal *TC* of their dielectric (Fig. 1–20). The *TC* becomes more steeply negative as the temperature approaches −55°C. (If, in your application, the *TC* is critical over some other range than +25°C to +85°C, Military Specification MIL C–20D and EIA Specification RS–198 provide the required additional information.) There is also a tolerance on the nominal *TC* that varies from ±30 ppm/°C for NPO to ±1000 ppm/°C for N5600. Effectively, an NPO ±30 ppm/°C ceramic capacitor will have a temperature coefficient falling somewhere between a +30 ppm/°C and −30 ppm/°C limit. The tolerance is greater for capacitance values below 10 pF, because the *TC* of stray capacitance becomes a major factor.

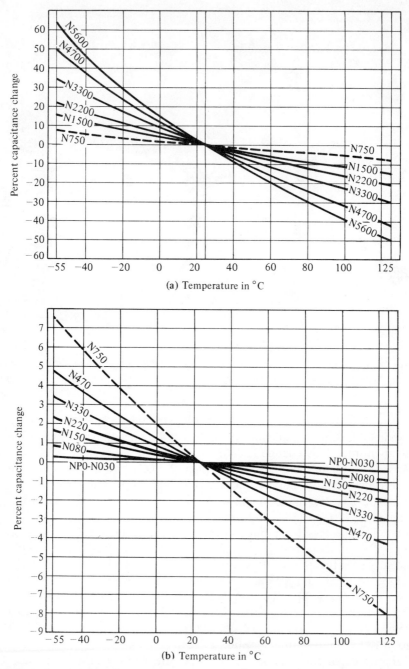

Fig. 1-20 Capacitance-temperature characteristics of ceramic temperature compensating capacitors **(a)** low range, **(b)** high range. The N750 curve is dashed to show relative scale. *(Courtesy of Erie Technological Products)*

Class I dielectrics from P100 to N1500 have a Q exceeding 1,000, and their voltage coefficient, frequency effects below 100 MHz, and aging are negligible. Class I dielectrics in the range of N2200 through N5600 are less stable and show some of the characteristics of Class II dielectrics.

These capacitors are widely used in tuned circuits, RC networks, and in any application where their accurate and predictable temperature coefficients can be used to compensate the temperature coefficients of other circuit components. Class I capacitors with low-range temperature coefficients are often used for their stability rather than for temperature compensation.

Class II high-K ceramic capacitors provide considerably more capacitance in the same package size as the Class I types, because of the higher dielectric constants of the ceramics used, but they are not as stable (Fig. 1–21). Capacitance varies with temperature, as do the dissipation factor and insulation resistance. Capacitance also varies with voltage (ac and dc have different effects), with frequency, and with aging.

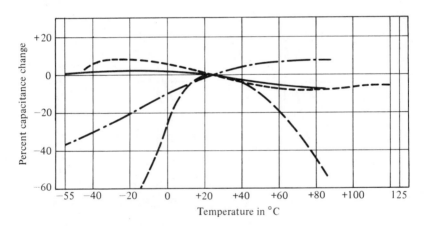

Fig. 1–21 Capacitance-temperature characteristics of typical Class II ceramic dielectrics.

The ceramic formulations used in Class I capacitors are not ferroelectric. This is important because it is the factor responsible for the temperature, voltage, and time stability characteristics of the capacitor.

Class II dielectrics are based on barium titanate. Modifiers, such as bismuth stannate, niobium, or tantalum pentoxide, may be added. These dielectric materials are ferroelectric and possess, in varying degrees, the characteristics for which high-K general application ceramic capacitors are known, including high capacitance, nonlinear temperature coefficients, ac and dc voltage sensitivity, and capacitance hysteresis. The principal advantage offered by these materials is the high dielectric constant, which ranges in value from 200 to as high as 16,000, depending on the mix used.

Barium titanate ceramic capacitors exhibit ferroelectric phenomena over the temperature range of $-55°C$ to $+125°C$. In the crystal structure of a ferroelectric ceramic dielectric material, various domains are present, and these domains have centers of positive and negative charges and can be considered as dipoles. The application of an electrical field changes the orientation of the dipoles. Since the dipoles change orientation at a rate generally slower than the change in the applied field, a hysteresis similar to that shown in Fig. 1–22 occurs. This hysteresis is important because the dielectric constant is dependent on the orientation of the dipole in relation to the charge.

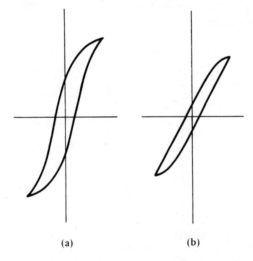

(a)   (b)

Fig. 1–22 Hysterysis loops for crystalline barium titante, **(a)** de-aged, **(b)** after one week. *(Courtesy of Sprague Electric Co.)*

Because of the ferroelectric effects, the dielectric constant and the capacitance, plus the dissipation factor, will change with frequency and voltage. These changes can be very significant.

Effects of Temperature. Although the various temperature effects must be considered individually, the parametric changes are combined algebraically.

The temperature at which the ceramic produces the highest dielectric constant, and above which the material becomes substantially nonferroelectric, is termed the Curie point. It is analogous to the Curie point in magnetism. This temperature is about 125°C for pure barium titanate, but the added modifiers both shift the Curie point over the temperature range and suppress it in some ranges, making possible a range of temperature characteristics (Fig. 1–21).

In addition to changing capacitance, a change in temperature also affects the dissipation factor (power factor) and insulation resistance. Many ceramic capacitors show a decrease in dissipation factor with an increase in temperature. This characteristic differs from many other dielectrics, but it is desirable because it adds stability in ac applications where heat dissipation is a problem.

The insulation resistance of the dielectric decreases with increasing temperature semilogarithmically. As the encasing material affects the overall insulation resistance, specification values for capacitors are composites. Most often, the insulation resistance is high enough that it is more important as an indicator of device quality, rather than as a design parameter, even at elevated temperatures.

Effects of Voltage.   The application of voltage to Class II ceramic capacitors produces a change in capacitance and power factor because of the ferroelectric phenomena. Both ac and dc voltage produce a change, but differently.

Low values of ac voltage (up to 20 volts) at 1 kHz produce an increase in both dissipation factor and capacitance. This is not usually a design problem. The change is actually dependent on the volts applied per mil of dielectric thickness and normally is minimal with devices such as discs, with dielectrics typically 10 mils to 20 mils thick. The changes become significant, however, with more sophisticated multilayer monolithic capacitors where 0.001-inch thick dielectrics are common.

A direct-current voltage produces a negative change in capacitance (it may go slightly positive with very low stress levels), which is greatly dependent on formulation, frequency of measurement, and voltage level. This change is often of extreme significance. See Fig. 1–23 for typical ac and dc voltage coefficient curves.

Voltage-Temperature Limits (TCVC).   The effects of voltage and temperature on the capacitance of ceramic capacitors may be added algebraically to get a combined voltage–temperature change. If the frequency is held constant, the maximum capacitance change due to the combination of a specified temperature excursion

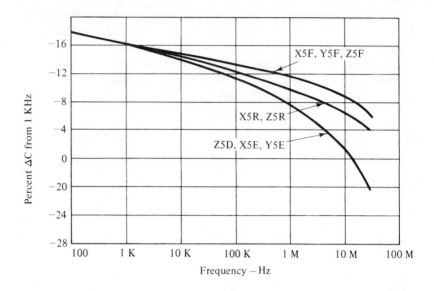

Fig. 1-23 Typical Class II ceramic dielectric ac and dc voltage coefficients. *(Courtesy of Erie Technological Products)*

and voltage is called the *voltage-temperature limit,* or TCVC. This is an important figure, since at a given frequency, almost all change is due to voltage and temperature. Note, however, that since the voltage coefficient changes with temperature, adding the voltage coefficient (determined at 25°C) to the temperature coefficient can be misleading.

Effects of Frequency. Class II barium titanate ceramic capacitors are frequency-sensitive (Fig. 1-24). The degree of sensitivity is dependent to a great extent upon the type of additives used to achieve temperature stability. Basically, the mechanism is one of dipole response to the alternating field. Some of the additives inhibit dipole reorientation at higher frequencies. The result of such inhibiting is a decrease in capacitance and an increase in dissipation factor.

Effects of Time (Aging). Another characteristic of barium titanate ceramic capacitors is their loss of capacitance with time. The aging rate is expressed as a negative percent per decade of time. It is associated with the ferroelectric state and is related to the geometry of the crystal structure.

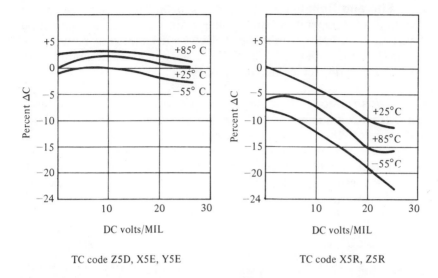

Fig. 1-24 Typical Class II dielectric capacitance-frequency characteristics. *(Courtesy of Erie Technological Products)*

The dissipation factor of these capacitors also decreases with age, but since a decrease here is desirable and usually insignificant anyway, it is not generally worried about.

The effect of aging in a capacitor application is of particular importance when the capacitance tolerance must be tight. In this circumstance, the aging rate may cause the capacitor to drift out of tolerance on the low side. Thus, it would be impractical to specify a 5% tolerance for a unit with a 2%/decade of time aging rate. Designing the capacitor with an initial value large enough to compensate for long aging will cause the units to be out of tolerance on the high side each time de-aging occurs. This is especially true of military systems where the high ambient operating temperatures of +125°C will, in many cases, cause some de-aging. For this reason, Class I ceramics should be used in tight-tolerance applications.

Voltage Aging. With Class II ceramic capacitors, there is also a voltage–aging effect. After the ceramic capacitor has been life-tested, there is normally a drop in capacitance, which is greater than would be attributed to the ordinary time-based aging.

The drop in capacitance is due to changes in the crystal structure caused by voltage, not time. The effect is more pronounced at higher temperatures, particularly if a high dc voltage has been applied.

Life and Reliability. Ceramic capacitor life is affected by the duration of temperature and voltage stress, once manufacturing faults such as voids within the dielectric, delamination between an electrode and a dielectric, and dielectric contamination are eliminated by burn-in, screening, and debugging.

Construction. Disc ceramic capacitors (Fig. 1–25) are fabricated from ceramic mixes with one metalized electrode on each surface (Fig. 1–26). The ceramic powder is mixed with a resin and compressed. Conductors are silver paint applied to both sides of the disc, usually by silk-screening. After the leads are attached by soldering, a conformal coating of a suitable resin is applied for environmental protection.

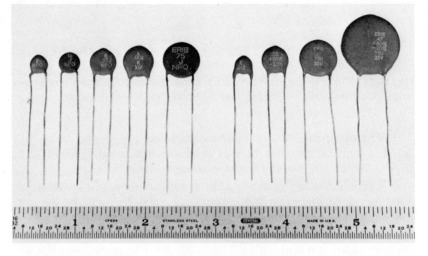

Fig. 1–25 Disc ceramic capacitors. Left: 500 dcwv types, values range from 1 pF to 4500 pF for temperature-compensating types, from 47 pF to 35,000 pF for general-purpose types. Right: 25 dcwv semiconductor high capacitance types, values range from 0.01 μF to 0.47 μF. *(Courtesy of Erie Technological Products)*

Tubular ceramic capacitors (Fig. 1–27) are formed from ceramic mixes extruded through a die to form the tube. One electrode is painted on the inner surface and terminated at one end of the tube, and the other electrode painted on the outer surface and terminated at the opposite end (Fig. 1–28). Silver, bonded with glass flux, is the most common electrode material, although others are also used. The capacitance of this style capacitor is generally limited by the configuration.

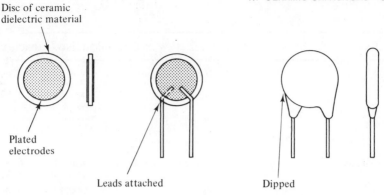

Disc of ceramic
dielectric material

Plated
electrodes

Leads attached   Dipped

Fig. 1–26   Disc ceramic capacitor construction.

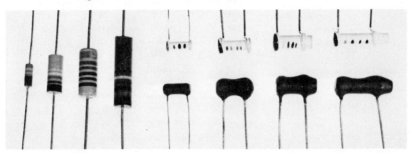

Fig. 1–27   Tubular ceramic capacitors. **Left:** Molded. **Top:** Enamel coated. **Bottom:** Dipped phenolic. Values range from 0.42 pF to about 1000 pF. Capacitors are shown full size. *(Courtesy of Erie Technological Products)*

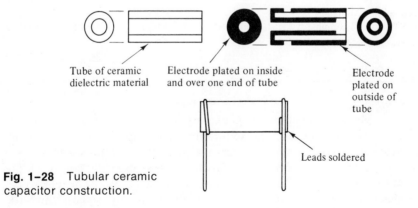

Tube of ceramic   Electrode plated on inside
dielectric material   and over one end of tube

Electrode
plated on
outside of
tube

Leads soldered

Fig. 1–28   Tubular ceramic
capacitor construction.

Ceramic disc capacitor varieties are described in Table 1–5 and ceramic tubular capacitors in Table 1–6.

TABLE 1-5   Summary Characteristics of Ceramic Disc Capacitors

	Class I	Class I	Class I	Class II
	NPO *Temperature* *Stable*	*N030–N750* *Temperature* *Compensating*	*N1500–N5600* *High-Range* *Temperature* *Compensating*	*Stable*
Working Voltage	6000 vdc	6000 vdc	6000 vdc	6000 vdc
	5000	5000	5000	5000
	4000	4000	4000	3000
	3000	3000	3000	2000
	2000	2000	2000	1000
	1000	1000	1000	600
	600	600	600	500
	500	500	500	100
	100	100	100	
Temperature Characteristic	NPO	N030	N1500	±3.3%
		N080	N2200	±4.7%
		N150	N3300	
		N220	N4200	
		N330	N4700	
		N470		
		N750		
Operating Temperature				
+10°C to +85 °C				x
−30°C to +85 °C				
−55°C to +85 °C	x	x	x	x
−55°C to +125°C				
−55°C to +150°C				
Capacitance Minimum	1.0 pF	1.0 pF	6.0 pF	
Maximum	390 pF	340 pF	3700 pF	4500 pF
Tolerances Available (above 10 pF)	±5%	±5%	±5%	
	±1%	±2%	±2%	
		±1%	±1%	

   Monolithic ceramic capacitors (Fig. 1–29) are rapidly replacing conventional ceramic capacitors in many applications. They are made of alternate layers of ceramic dielectric materials—mixtures of fine ceramic powders and resins—and silk-screened layers of precious metal paste to form electrodes (Fig. 1–30). The electroded sheets are stacked and compacted, then cut into squares and rectangles. Each (multilayer) piece contains several electrodes which extend alternately to each of the cut ends.

   The units are fired to remove organic materials and then sin-

TABLE 1-5   Continued

Class II	Class II	Class I	Semiconductor Type			
Semi-stable	General-Purpose	AC (UL-listed)	Low-Voltage Stable	Capacitor with Spark Gap	Low-Voltage Semi-stable	Low-Voltage General Range
6000 vdc	7500 vdc	125 vdc	25 vdc	3000 vdc	100 vdc	75 vdc
5000	6000	(1,400 vdc)	18	1500	50	50
3000	5000		16	1000	25	3
2000	4000		12		20	
1000	3000		3		16	
500	2000				12	
250	1000				10	
100	600					
	500					
	100					
±10%	+22–33%	N330±500 ppm/°C	±4.7%	±10%	±7.5%	
±7.5%	+22–56%	N3300±2500 ppm/°C	±7.5%	+22–56%		+22–33%
	+22–82%					
	+22–90%					
x	x		x	x		
x	x		x			
x	x	x	x		x	x
	x					
	x					
470 pF	1.5 pF·	470 pF	0.01%	75 pF	0.005 μF	0.05 μF
0.02 μF	0.05 μF	0.5 μF	2.2 μF	0.1 μF	0.47 μF	2.2 μF
±10%	±20%	±20%	±5%	±20%	±20%	+100–0%
		±10%		±10%		±80–20%
						±80–30%

tered at 1200°C to 1450°C. Ends with exposed electrodes are dipped into a metal paste (generally silver or palladium-silver) and fired at about 750°C to form solderable terminations.

At this stage, the monolithic capacitor is a chip—and is sold in this chip form as a hybrid-circuit component. For general circuit applications, the chip capacitors are encased in a variety of ways for environmental protection and ease of handling: Cylindrical or rectangular molded cases with either axial leads or radial leads, or a single-ended configuration for PC board mounting; hermetically

TABLE 1-6 Summary Characteristics of Tubular Ceramic Capacitors

	Class I			Class II		
	Temperature Stable	Temperature Compensating	High-Range Temperature Compensating	Stable	Semistable	General-Purpose
Working Voltage	500 vdc 200 100	500 vdc 200 100	500 vdc 200 100	1000 vdc 500 200 100	1000 vdc 500 200 100	1000 vdc 500 200 100
Temperature Characteristic	P100 P030 NP0 N030 N080	N150 N220 N330 N470 N750	N1500 N2200 N3300 N4200 N4700 N5600	±3.3% ±4.7%	±7.5% ±10% ±15% ±22%	+22–33% +22–66% +22–82% +22–90%
Operating Temperature						
+10°C to +85 °C						x
-30°C to +85 °C				x	x	x
-55°C to +85 °C				x	x	x
-55°C to +125°C	x	x			x	x
-55°C to +150°C			x			x
Capacitance Minimum Maximum	0.42 pF 68 pF	0.63 pF 137 pF	1.5 pF 1100 pF	– 1800 pF	– 1800 pF	– 5600 pF
Tolerances (above 10 pF) Available	±20% ±10% ±5 % ±3 %	±20% ±10% ±5 % ±3 %	±20% ±10% ±5 % ±3 %	±5%	±10%	±20% ±10%

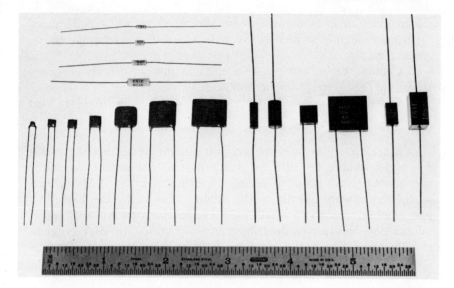

Fig. 1-29 Monolithic ceramic capacitors. **Top left:** Glass sealed. **Left:** Dipped polymer coated. **Right:** epoxy molded. Values range up to 0.47 µF at 100 dcwv. *(Courtesy of Erie Technological Products)*

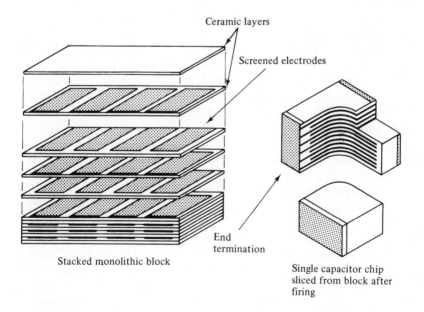

Ceramic layers

Screened electrodes

End termination

Stacked monolithic block

Single capacitor chip sliced from block after firing

Fig. 1-30 Monolithic ceramic capacitor construction.

sealed glass tubes with axial leads; enclosed in premolded epoxy cases subsequently sealed with epoxy; radial-lead conformal epoxy coating, etc.

Ceramic capacitors are also made in button, cartwheel, doorknob, and feed-through configurations.

Ceramic monolithic capacitors are described in Table 1–7 on pages 42 and 43.

Marking. Part marking and part numbering are not the same, which is fortunate, because the standard industrial ceramic capacitor part number runs to as many as 16 characters and digits, which would be awkward to print on the smaller sizes to say the least. The part number, which you would use to order the part, contains coded information completely describing the capacitor. An EIA-assigned manufacturer's symbol is often the only marking for which there is space on the smaller parts.

1.8 ALUMINUM ELECTROLYTIC CAPACITORS

These capacitors (Fig. 1–31) are electrochemical devices. Two aluminum foils, separated by insulating papers, are wound convolutely into a roll or cylinder. The roll is impregnated with a liquid electrolyte, electrically aged and stabilized, and then sealed in an aluminum container.

The aluminum electrolytic capacitor is not exactly an ideal capacitor. Direct-current leakage is high, and inherent internal inductances drastically limit high-frequency performance. Tolerance on nominal capacitance can run as loose as -10% to $+150\%$, and capacitance drops with age. Low-temperature stability is poor; they sometimes leak chemicals and have been known to burst open when overloaded. Nevertheless they are widely used. Capacitances range upward to one farad, they are lightweight, and they are inexpensive.

Construction. In the polarized electrolytic capacitor (Fig. 1–32), one of the aluminum foils is the anode plate. An oxide coating formed on the anode foil is the dielectric. The electrolyte itself is the cathode. The cathode foil merely serves as the connection between the electrolyte and the external circuit. The paper separators prevent the possibility of shorts between the anode and cathode foils, hold the electrolyte in close contact with the foils, and provide a uniform distribution of electrolyte. In nonpolar capacitors, both aluminum foils have equal oxide formations. Actually, the nonpolarized capacitor is two polarized electrolytic capacitors back-to-back.

Fig. 1–31 Aluminum electrolytic capacitors. Shown: Computer grade, twist mount, small axial lead, PC board mounting types. *(Courtesy of Mallory Capacitor Co.)*

Foils may be either plain or etched. Electrochemical etching produces a microscopically rough surface, which can increase the effective electrode area as much as 50 to 1 with a proportional increase in obtainable capacitance. The plain or etched foil is then passed through an electrolyte solution tank with voltage applied across the foil and solution. The thickness of the barrier oxide layer formed on the surface of the foil is proportional to the voltage. The working voltage of the capacitor will be proportional and capacitance will be inversely proportional to the thickness of the film. The quality and thickness of the formed dielectric also determines the dielectric loss in the capacitor, expressed as dc leakage current.

Generally, aluminum electrolytic capacitors have a maximum operating temperature of 85°C; some are rated at 105°C or 125°C.

TABLE 1-7 Summary Characteristics of Monolythic Ceramic Capacitors

| | Temperature Compensating | High-Range Temperature Compensating | Flat Plates | | | |
			High-K	Ultra-stable	Stable	General-Purpose
Working Voltage	200 vdc 100	200 vdc 100	200 vdc 100	500 vdc 200 100 50	500 vdc 200 100 50	500 vdc 200 100 50 25
Temperature Characteristic	NP0 N030 N080 N150 N220 N330 N470 N750	N1500 N2200 N3300 N4200 N4700 N5600	±3.7% ±4.7% ±7.5% ±10 % ±15 % ±22 % +22–33% +22–56% +22–82%	0±30 ppm/°C	±15%	+22–82% +22–56% +22–33% +15–25% +15–40%
Operating Temperature						
+10°C to +85 °C			x			
−30°C to +85 °C			x			
−55°C to +85 °C			x			x
−55°C to +125°C	x		x	x	x	x
−55°C to +150°C		x				
Capacitance Minimum	1.0 pF	6.8 pF	2 pF	4.7 pF	470 pF	270 pF
Maximum	850 pF	8600 pF	0.059 µF	0.01 µF	0.56 µF	6.8 µF
Tolerances (above 10 pF) Available	±1% (N750, ±2%)	±2% (N1500, N2200) ±5%	±20% ±10% ±5 %	±10% ±1 %	±10% ±5 %	±20% ±10% ±5 %

TABLE 1-7   Continued

	Tubular				Rectangular Molded		
	Ultra-stable	Stable	Semi-stable	General-Purpose	Ultra-stable	Stable	Semi-stable
Working Voltage	100 vdc	100 vdc	100 vdc 50	100 vdc 50	100 vdc 50	200 vdc 100 50 25	200 vdc 100 50 25
Temperature Characteristic	0±30 ppm/°C 0±25 ppm/°C	±15%	+15%	+22-33% +22-50% +22-82% +15-25% +15-40%	0+60 ppm/°C	±15%	+22-55%
Operating Temperature							
+10°C to +85 °C	x						
−30°C to +85 °C							x
−55°C to +85 °C				x	x		
−55°C to +125°C		x	x			x	
−55°C to +150°C					x		
Capacitance Minimum Maximum	0.1 pF 1000 pF	10 pF 0.1 µF	10 pF 0.27 µF	1000 pF 0.47 µF	1200 pF 0.33 µF	1500 pF 2.2 µF	.033 µF 2.2 µF
Tolerances (above 10 pF) **Available**	±10%	±10%	±20% ±10%	±20% ±80-20% ±10% ±5 %	±20% ±10% ±5 %	±10%	±20% +80-20%

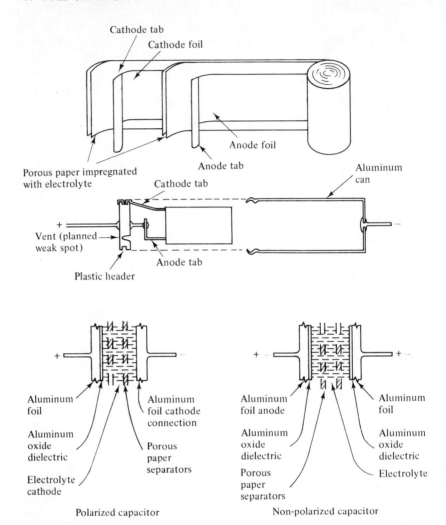

Fig. 1–32   Aluminum electrolytic capacitor construction.

The choice of electrolyte to a large extent determines the temperature range of operation, the dissipation factor, the magnitude of the dc leakage current, the capacitance stability, the ac ripple rating, and the life expectancy of the capacitor. Since aluminum is attacked by water, especially at high temperatures, all traces of water must be eliminated. The electrolyte must be totally unaffected by aluminum or aluminum oxides. Conductivity must be as high as possible without the sparking phenomena known as *scintillation*. Because the aluminum oxide probably requires repair now and then during its life span, the electrolyte must also have the ability to form oxide films on the foils. The electrolyte should also be able to recombine any gases produced by leakage currents into ionic components to avoid build-up of gas pressure inside the can. A vent is usually provided for gas escape in the event of abnormal internal pressure build-up.

Operating an aluminum electrolytic capacitor at temperatures beyond the range of the electrolyte results in sharply reduced capacitance at low temperatures and short life at high temperatures.

For applications where the ripple current exceeds the limits for polarized electrolytic capacitors, semipolarized types can be used. In these, the anode has a relatively thick oxide formed, and the cathode foil connection becomes a secondary anode with a thin, oxide layer. These types usually have to be specially ordered.

While the use of polarized electrolytic capacitors is restricted to circuits having net current flow in one direction only, nonpolarized electrolytic capacitors can be used in audio networks and ac motor-starting applications.

Aluminum electrolytic capacitors are sometimes constructed with more than one capacitor section in the same can. As the cathode is almost always connected to common ground, several capacitor anodes may be included in one convolute roll and can. The capacitor sections can be operated at different voltages or at the same voltage. Each capacitor section will have its own anode formed according to the working voltage desired for that section. Since the electrolyte is common to all sections, the highest voltage rated section will determine the type of electrolyte.

Electrolytic capacitors usually fail in one of three modes: They develop an early-failure short-circuit or open-circuit, usually due to a manufacturing defect; the vent may erupt due to overvoltage or excessive temperature; or there is gradual deterioration in capacitance loss, dissipation factor rise, increased leakage current, and progressive material failures usually ending in loss of the electrolyte through the seals.

Twist-Mount Aluminum Electrolytic Capacitors. These self-mounting capacitors are widely used in radio and television receivers, consumer audio equipment, commercial electronic and broadcast equipment, and industrial electronic equipment. They are made in three styles, the difference being the mounting arrangement:

1. The chassis style is mounted on the metal chassis by means of integral mounting ears which fit in slots punched in

the chassis, or more usually by means of a metal or phenolic mounting plate fitting over a hole in the chassis. The capacitor is secured in place by twisting the ears. They may also be mounted below the chassis in spring mounting clips.

2. The PC board style mounts with deformed terminal tabs that hold the capacitor in position until it is dip soldered. These terminals also have shoulders to hold the capacitor body above the board surface. These terminals are not on standard PC board spacing and require oriented rectangular holes.

3. The third style has straight mounting terminals dimensioned for making wire wrap connections. They can also be used for dip soldering.

All three varieties are supplied in bare aluminum cans, or with cardboard tube or plastic film outer insulation for applications where the capacitor cathode and can will not be at chassis ground potential.

Twist mount capacitors are generally made in two can diameters, 1 inch and 1 3/8 inch in mounted heights ranging from 2 to 4 inches, with up to three sections in the 1-inch can, and up to four sections in the 1 3/8-inch can.

In the vacuum tube era, distributor catalogs had page after page of listings for standard twist mount capacitors. Today, most are manufactured in voltage and capacitance combinations to meet specific high-volume equipment requirements. Many combinations having wide use in laboratories, general circuit development, and replacement are made as standard units (chassis mount only) and stocked by distributors.

General-Purpose Tubular Aluminum Electrolytic Capacitors. These low-cost capacitors are used in radio and television receivers and consumer audio equipment for power supply filtering, bypass and audio coupling applications. They are commonly used in point-to-point wiring supported only by their leads; the capacitors are also used with mounting straps.

These capacitors are sealed in tubular aluminum cases using, typically, a rubber-phenolic end disc with the can securely spun over the elastomer so that the capacitor will not leak out, dry out, or be contaminated with atmospheric moisture. Some types have a vented construction. The capacitors are made in single or dual sections or in isolated dual sections. Lead wires are axial; bare or insulated; some types have all lead wires at one end; some types are made with solder tabs instead of wires. These general-purpose capacitors are also made in nonpolarized versions.

Miniature Aluminum Electrolytic Capacitors. These miniature polarized aluminum electrolytic capacitors are "miniature" only when compared to the can size for equivalent capacitance and voltage ratings of a few decades ago. Advances in aluminum capacitor technology have wiped out the size advantage of tantalum foil capacitors; aluminum capacitors also cost a lot less and weigh less. These capacitors are made in axial-lead and single-ended styles and are used in a wide variety of applications including radio and television receivers, audio amplifiers, business machines, control and test equipment, and computers.

Computer-Grade Aluminum Electrolytic Capacitors. The term *computer grade* is loosely used to define high-quality, very high-capacitance aluminum electrolytic capacitors such as those used in the filter circuits of high-current dc power supplies for digital computers, and laboratory, military, and aerospace electronic equipment. They are available in capacitance values up to *one farad*.

The capacitors are packaged in drawn aluminum cans with molded headers. Terminals are either threaded inserts or solder lugs. The tapped inserts are preferred where strap or bus bar connections are used. Typically, the seal consists of crimping the beaded aluminum onto a rubber gasket recessed in the rigid molded cover. Pressure-sensitive safety vents made of rubber are used on all case covers. The capacitors are supplied either with bare cans or with an outer plastic sleeve.

These capacitors have an operating life of up to 20 years, low ESR, and low impedance, and can withstand ripple currents up to 30 amperes, depending on rating.

For more modest capacitance requirements—say 12,000 μF at 10 wvdc instead of 390,000 μF—miniature (relatively speaking) axial-lead computer-grade capacitors are available. These units are normally furnished as polarized capacitors with the cathode connected to the can; nonpolarized capacitors can be obtained on special order. Table 1–8 describes aluminum electrolytic capacitor variations.

1.9 TANTALUM ELECTROLYTIC CAPACITORS

Despite a considerably higher cost than aluminum electrolytic capacitors, tantalums (Fig. 1–33) are extensively used in miniaturized equipment and computers because of their smaller size, stability, wide operating temperature range and long reliable operating life.

TABLE 1-8   Summary Characteristics of Aluminum Electrolytic Capacitors†

	General Purpose			Computer Grade	
	Twist Mount	Tubular	Miniature Tubular	Standard	Axial-Lead
Maximum Can Volume	6.5 cu in.	6.2 cu in.	0.300 in. max.†	61 cu in.	3 cu in. max.†
Capacitance, Minimum	10 μF	1 μF	1 μF	15 μF	1 μF
Maximum 3vdc	–	50,000 μF	2200 μF	570,000 μF	18,000 μF
Capacitance 6	2000 μF	40,000	1500	480,000	12,000
of WVDC 10/12	–	30,000	800	360,000	–
15/16	4000	20,000	800	270,000	8000
25	1000	13,000	500	160,000	4700
50	1500	8000	300	70,000	2300
150	500	1000	40	13,000	530
250	200	600	15	4500	300
350	250	300	12	2500	160
450	150	200	8	1200	75

† Typical maximum capacitance for working voltage, polarized capacitors, +85°C operation.

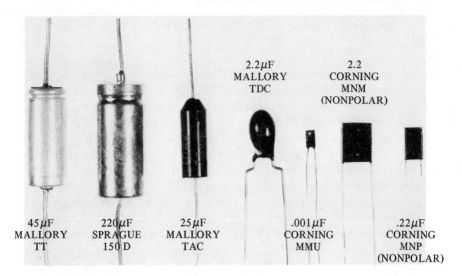

Fig. 1–33 Tantalum electrolytic capacitors. Aluminum electrolytic capacitor at left for size comparison. Capacitance values are maximums at 20 v dcwv for the case size. Capacitors are shown full size. *(Courtesy of Corning Glass Works, Mallory Capacitor Co, Sprague Electric Co.)*

Tantalum capacitors are made in plain and etched foil types, polarized and nonpolarized. They are also made in a "wet slug" sintered anode style and a dry-electrolyte sintered anode style. The latter presents no sealing problems.

Tantalum-Foil Electrolytic Capacitors. These capacitors have no advantage over aluminum electrolytics in capacitance to volume ratio; in fact, they are bigger than some types and cost considerably more. They do have advantages in that they are not susceptible to halogenated hydrocarbon contamination and they are available in hermetically-sealed cases. Construction is similar to aluminum foil capacitors, but the materials are different. Tantalum foil is used in place of aluminum foil. The capacitors are made in both plain and etched versions, both polarized and nonpolarized. Paper is used for a separator the same as in an aluminum electrolytic capacitor. Etched versions offer a considerable advantage in capacitance-to-volume ratio, however, over plain foil tantalums. Various electrolytes are used depending on the desired voltage rating. Cases are made of silver-plated brass, stainless steel, aluminum, copper, or titanium.

Tantalum is the most stable of all anodic film-forming materials; reforming is unnecessary, and the capacitors have a long shelf life.

Wet-Slug Tantalum Capacitors. These miniature and sub-miniature polarized capacitors are one-fourth the size of aluminum capacitors of comparable capacitance, temperature, and voltage rating. Stability is good over temperature extremes, and the capacitors have a long service life. These capacitors have the highest volumetric efficiency of any capacitor. Their small size makes them particularly useful in printed circuits, with DIP integrated circuits, and with other small module type packages.

In these capacitors (Fig. 1–34), the anode is made from pure powdered tantalum, which is pressed into a cylindrical pellet and then fired in a 1600°C to 2000°C vacuum furnace fusing the minute tantalum particles together without melting them into a solid mass. The lead wire can be welded on at this time, or it can be pressed in place at the time the pellet is made. The porous pellet is then placed in an anodic bath where an oxide coating is formed on the entire spongy surface, which is considerable in comparison with the surface of a solid cylinder of the same size. The cathode is the electrolyte and the specially prepared inside surface of the metal case. The electrolyte, which must have high conductance, is either a liquid or gelled sulfuric acid or lithium chloride. Sealing the case is critical; no one wants acid oozing around their circuitry. Several seals are used: Elastomers, combinations of Teflon ® and elastomer seals, and glass-to-tantalum hermetic seals. A major problem of wet tantalums has been their inability to withstand reverse voltage. A recent development has been the tantalum case, in which the entire capacitor is tantalum, doing away with problems such as silver migration and the reverse-voltage cell effect. These all-tantalum capacitors have the inherent ability to withstand 3 volts reverse voltage.

Solid Electrolyte Tantalum Capacitors. The solid electrolyte capacitor is quite similar to the wet type in construction. The anode is a sintered tantalum pellet on which is formed a tantalum pentoxide film, which is the dielectric. The oxide film is formed on the surface of the pellet particles by immersing the pellet in a liquid electrolyte and applying a voltage. This converts the surface of the wetted tantalum to tantalum pentoxide. The pellet is next immersed in a solution of manganous nitrate and then heated in an oven to convert the manganous nitrate to manganese dioxide. A carbon and metal film applied over the electrolyte is the cathode. Over that, a copper or silver coating provides a solderable surface. The solid

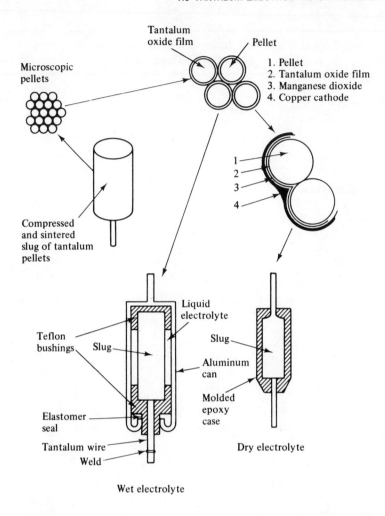

Fig. 1–34 Sintered-anode tantalum capacitor construction.

electrolyte eliminates problems with electrolysis and leakage, two life-limiting factors. The characteristics of the solid electrolyte capacitor differ significantly from the wet variety. The capacitance/frequency, capacitance/temperature characteristics, and dissipation factor are improved over other electrolytic capacitors that have liquid or aqueous electrolytes.

These capacitors are packaged in a variety of styles. Physical size, standard capacitance value ranges and dc working voltages for all types of tantalum capacitors are compared in Table 1–9.

TABLE 1-9 Summary Characteristics of Tantalum Electrolytic Capacitors

| | | Foil | | | | Sintered Anode Liquid or Gelled | |
| | | Plain | | Etched | | | |
		Polar	Nonpolar	Polar	Nonpolar	Miniature	Standard
Tolerances, Standard				−15 +30% −15 +50% −15 +75% ±20% ±15%		±20% ±10%	±20% ±10%
Maximum Case Volume (arbitrary)				0.6 cu. in.		0.03 cu. in.	0.15 cu. in.
Minimum Capacitance		0.25 μF	0.15 μF	1.0 μF	1.0 μF	1.0 μF	1.7 μF
Maximum Capacitance each WVDC	2 vdc	−	−	−	−	−	−
	3	1300	−	−	−	−	−
	4	−	−	−	−	−	−
	6	560	470	−	−	470	1200
	10	420	280	−	−	330	750
	15	320	200	2500	2000	220	540
	20	−	−	−	−	180	470
	25	210	120	1600	1100	−	−
	30	180	100	1400	850	120	300
	35	−	−	−	−	−	−
	50	120	65	580	310	50	160
	60	100	53	−	−	56	140
	75	88	47	300	50	47	110
	100	70	37	210	35	−	86
	125	−	−	−	−	−	56
	150	47	24	120	18	−	47
	200	35	17	66	−	−	−
	250	30	15	45	−	−	−
	300	18	−	28	−	−	−

1.10 ALTERNATING-CURRENT OIL CAPACITORS

These capacitors (Fig. 1–35) are designed primarily to provide large capacitance for industrial ac applications and to withstand the large currents and high peak voltages encountered at power line frequencies. Applications include phase-splitting, power factor correction, voltage regulation, control equipment, stabilizing transformer

TABLE 1-9   *Continued*

"Wet Slug" Electrolyte	Sintered Anode Dry Electrolyte					
High-Range Standard	Ultra-miniature Dipped	Sub-miniature Molded PC Board	Sub-miniature Molded & Tubular	Miniature Molded	Standard Dipped	Standard Tubular
±20% ±10%	±20%	±20%	±20% ±10%	±20% ±10% ±40%–20%	±20% ±10%	±20% ±10%
0.15 cu. in.	0.003 cu. in.	0.012 cu. in.	.01 cu. in.	0.03 cu. in.	0.10 cu. in.	0.08 cu. in.
1.7 μF	0.1 μF	0.1 μF	0.08 μF	0.001 μF	0.10 μF	.0047 μF
—	10.0	—	68	22.0	—	—
—	6.8	100	—	—	—	—
—	4.7	—	68	10.0	330	—
2200	3.3	68	56	6.8	330	1000
1500	2.2	47	39	4.7	220	560
1000	1.5	27	22	3.3	150	330
—	1.0	22	15	2.2	100	220
—	—	—	10	—	68	—
560	—	—	—	—	—	100
—	4.7	10	6.8	—	47	68
330	—	—	4.7	—	22	39
270	—	—	—	—	—	—
220	—	—	—	—	—	—
86	—	—	—	—	—	—
56	—	—	—	—	—	—
39	—	—	—	—	—	—
—	—	—	—	—	—	—
—	—	—	—	—	—	—
—	—	—	—	—	—	—

circuitry, fluorescent lamp ballasts, and ac motor-starting and running.

Construction. The construction of these capacitors consists of metal-foil electrodes separated by oil-impregnated kraft capacitor paper, kraft paper and film, or just film wound in a roll and encased in a hermetically-sealed metal housing. For ac applications, poly-

Fig. 1–35 Drawn steel cased ac oil-filled capacitors. These Sprague type 500P and 501P capacitors utilize a non-toxic biodegradable dielectric. Capacitance values range from 1 μF to 55 μF with ac working voltages ranging from 330 to 660 vac, 60 Hz. *(Courtesy of Sprague Electric Co.)*

propylene is the film used. The impregnant, usually a nonflammable, synthetic liquid or halogenated hydrocarbon (askarel), increases the dielectric constant of the paper, reduces corona, and increases dielectric strength.

Use of paper capacitors impregnated with a polychlorinated biphenyl (PCB) askarel oil is coming under pressure because of its toxic and persistent environmentally-hazardous qualities. Restrictions on the general use of PCBs in the food processing industry went into effect in the United States in 1973, and several nations, notably Japan, have banned import of this material in any form.

For such applications, oil capacitors are made with biodegradable impregnants, which are either silicones or blends of organic ester resins. However, these nonpolluting impregnants produce electrical characteristics different from PCBs, and they are flammable.

Volt–Ampere Rating. Operating temperature and applied voltage are the two major factors determining the life of an ac capacitor. Operating temperature is a combination of ambient temperature and the rise due to the power dissipated by the current passing through the capacitor. This current will vary with frequency.

This has produced a convenient volt–ampere rating system for ac capacitors.

$VA = 2\pi f C E^2 \times 10^{-6}$

$f =$ frequency in Hz

$c =$ capacitance in μF

$E =$ rated voltage   (Eq. 1–1)

The VA rating provides a basis for configuration design and for the determining of surface area of the container required to dissipate heat efficiently. In equipment, the capacitor should be located away from heat-producing transformers and high-power resistors, etc.

1.11 DIRECT-CURRENT OIL CAPACITORS

DC oil capacitors (Fig. 1–36) are designed for filtering, bypass, coupling, voltage doubling, arc suppression, and similar dc applications.

Fig. 1–36 High voltage, high temperature dual-dielectric oil-filled capacitor. For this style capacitor, ratings range from 8 μF at 5 kv dc to 0.5 μF at 50 kv dc. *(Courtesy of Cornell-Dubilier Capacitor Co.)*

They are made in a broad range of case styles, capacitance values and tolerances, and operating temperatures. Working voltage ratings run from 50 volts to 200,000 volts dc.

Construction. Direct current oil capacitors usually have kraft capacitor paper as a dielectric material, either alone or in combination with polyester film. The electrodes are metal foil. Various impregnants such as askarel, mineral oil, silicones and polyisobutylene are used to obtain desired capacitor characteristics.

Derating. Whenever there is an alternating component to the direct voltage, its peak value must be added to the direct voltage to obtain the operating voltage and should not be allowed to exceed the values in Fig. 1–37.

It is important to note that pulsing or rapidly discharging a general-purpose dc oil capacitor will materially shorten its life.

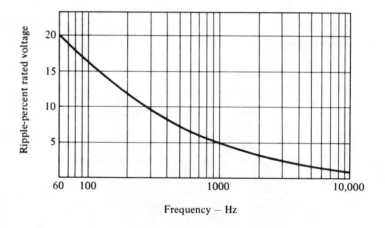

Fig. 1–37 DC oil-filled capacitor operating voltage derating for ripple.

1.12 ENERGY DISCHARGE CAPACITORS

High-voltage energy discharge capacitors (Fig. 1–38) provide a convenient method for storing electrical energy which can be released as needed for short periods of time (shots) under controlled conditions.

These capacitors are designed specifically for intermittent-duty discharge applications. Conventional dc capacitors operate at a relatively steady dc voltage; energy storage capacitors can be charged in

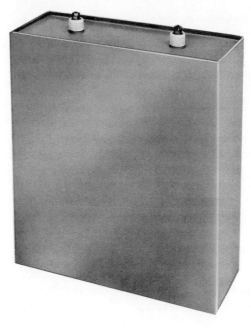

Fig. 1–38 Energy discharge capacitor designed for photocopier applications. Capacitor shown one-sixth size, has rating of 40 μF with a charging voltage of 3000 volts. *(Courtesy of Cornell-Dubilier Capacitor Co.)*

milliseconds and then discharged in time periods ranging from nanoseconds, seconds, minutes to hours. Very high instantaneous power levels are obtained in the energy discharge.

Light-duty applications for energy discharge capacitors include biomedical equipment, welders, high-intensity flash tubes, masers, lasers, etc. In these applications, the capacitors discharge into a resistive or other critically damped load. The stored energy is released over a relatively long time period with little or no voltage reversal. Heavy-duty applications may involve oscillatory discharges with voltage reversals of up to 90%.

Construction.   Electrodes are of high-purity aluminum foil in an extended foil (noninductive) construction. Where peak current and internal inductance demands are not severe, multiple-tab construction may be used; otherwise, extended foil construction is used.

The dielectric normally consists of multiple layers of kraft capacitor paper impregnated with electrical-grade castor oil.

Large capacitors have heavy gauge welded seam steel cases;

smaller capacitors may be supplied in lighter gauge drawn steel cans.

Terminal designs vary according to inductance and voltage requirements. Low-profile molded epoxy terminal headers provide low inductance and will withstand high voltage. Where inductance requirements are less severe, conventional ceramic solder-seal pillar bushing terminations are used.

Stored Energy. Energy discharge capacitors may be used individually or in banks ranging up to several million joules. The energy stored in joules (or watt-seconds) in a capacitor, or in a bank of capacitors, is given by

$$\text{Joules} = 1/2\ CE^2 \times 10^{-6}$$

Where:

C = total capacitance in μF

E = potential to which capacitor is
charged in volts   (Eq. 1–2)

Discharge Cycle. Discharging the capacitor into a low-resistance circuit produces a damped oscillatory current waveform (Fig. 1–39). To obtain maximum peak current, circuit inductance must also be minimized. Capacitor equivalent series resistance (ESR) and inductance must be considered. Banks of small capacitors instead of one large one can reduce inductance but at an increased cost.

The life of the capacitors is adversely affected by the peak in-

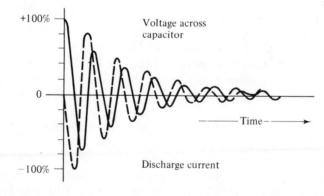

Fig. 1–39 Discharge wave form for capacitor discharge into a low-resistance circuit.

verse voltage occurring on the first reversal. Increasing circuit resistance will limit this voltage, but peak current is then also limited.

The life of energy storage capacitors is usually given in charge-discharge cycles (shots) or total time on charge. To meet size and cost requirements, some may be designed with a life of only 100 shots, but the specification can run into the thousands and millions.

In the design of these capacitors, provision must be simultaneously made to carry the high discharge current and to minimize self-inductance.

Table 1–10 describes the major varieties of these types of oil-filled capacitors. (See pages 60 and 61)

1.13 WHO MAKES WHAT

Table 1–11 lists some of the major capacitor vendors and the types of capacitors they make. (See page 62)

SUGGESTED READINGS

Alwitt, Robert S., and Reginald G. Hills, *The Chemistry of Failure of Aluminum Electrolytic Capacitors,* TP–64–11. Sprague Electric Co., North Adams, Mass., December 1964.

Burton, John H., and Edmund A. Bolton, *A Better Understanding of Ceramic Capacitors,* TP–67–15A. Sprague Electric Company, North Adams, Mass., 1967.

Dummer, G. W. A., and Harold M. Nordenberg, *Fixed and Variable Capacitors.* McGraw-Hill Book Co., New York, N. Y., 1960, pp. 77–191.

Henry, Edward C., *Electronic Ceramics.* Doubleday & Company, Inc., Garden City, N. Y., 1969.

McManus, Ronald P., *Aluminum Electrolytic Capacitors,* TP–72–2A. Sprague Electric Co., North Adams, Mass., 1972.

Moynihan, John D., *Electrolytic Capacitors in Timing Circuits,* TP–68–3. Sprague Electric Co., North Adams, Mass., 1968.

Moynihan, John D., *Non-Polar Electrolytic Capacitors in Timing Circuits,* TP–68–7. Sprague Electric Co., North Adams, Mass., 1968.

TABLE 1-10   Summary Characteristics of Oil-Filled Capacitors

	AC (60 Hz)			
	General-Purpose		Nonpolluting	
Case Style	Drawn steel case		Drawn steel case	
Capacitance, Minimum	1 μF		0.5 μF	
Capacitance of Working	(vrms)		(vrms)	
Voltage (Maximum)	165	50 μF	—	—
(other voltages available)	236	50	—	—
	330	90	330	45 μF
	370	90	370	50
	440	60	440	55
	660	25	660	25
Tolerance, Standard Available	±10%, ±6%, ±5%		±10%, ±6%	
Operating Temperatures (depend on impregnant)	−30°C to +55°C		−55°C to +70°C −40°C to +70°C	
Dissipation Factor	0.6%		0.5%	

	DC					
	Small		Standard		High-Voltage	
Case Style	Drawn steel, bathtub, and rectangular		Drawn steel		Welded steel	
Capacitance, Minimum	0.05 μF		0.5 μF		0.25 μF	
Capacitance of Working	(vdc)		(vdc)		(vdc)	
Voltage (Maximum)	100	4.0 μF	400	30 μF	5000	50 μF
(other voltages available)	200	2.0	600	35	7500	20
	400	1.0	1000	25	10,000	10
	600	2.0	1500	12	15,000	5.0
	1000	1.0	2000	10	20,000	2.0
			2500	7.5	25,000	2.0
			3000	5.0	30,000	1.0
					50,000	0.5
Tolerance, Standard Available	±10%		±10%		±10%	
Operating Temperatures (depend on impregnant)	−55°C to +85 °C −55°C to +125°C		−55°C to +85°C		-55°C to +40°C	
Dissipation Factor	1.0%		1.0%		1.0%	

TABLE 1-10   *Continued*

	Energy Discharge		AC/DC
	Light Duty	*Heavy Duty*	*General-Purpose*
Case Style	Drawn steel	Drawn steel, or welded cases, double-roll sealed	Drawn steel
Capacitance, Minimum	10 μF	10 μF	0.1 μF
Capacitance of Working Voltage (Maximum) (other voltages available)	(vdc) 2000 200 μF 2500 200 4000 200	(vdc) 2000 100 μF 2500 340 3000 260 4000 140 5000 90 6000 60 10,000 20	(vdc/vrms) 600/330 20 μF 1000/440 20 1500/660 15 2000/700 10 2500/800 8.0 3000/950 4.0
Tolerance, Standard Available	+20% −10%,	+20% −10%,	±10%
Operating Temperatures (depend on impregnant)	0°C to +40°C	0°C to +40°C	−55°C to +55°C (ac) −55°C to +85°C (dc)
Dissipation Factor	−	−	−

CATALOGS

Centralab Electronics Division, Globe Union, Inc., Milwaukee, Wisc.

Cornell-Dubilier Electric Corp. (CDE), Newark, N. J.

Corning Glass Works, Corning, N. Y.

Erie Technological Products, Inc., Erie, Pa.

Mallory Capacitor Co., Indianapolis, Ind.

Sangamo Electric Co., Pickens, S. C.

Sprague Electric Co., North Adams, Mass.

TABLE 1-11   Capacitor Manufacturers

	Arco Electronics	Aerovox	Cornell-Dubilier	Centralab	Corning	Erie	General Electric	Mallory	Nytronics	Sangamo	Sprague	TRW Capacitors
Paper & Film												
Metalized Paper											X	
Polyester	X		X				X				X	X
Metalized Polyester	X		X				X	X			X	X
Polyester and Paper			X								X	
Polystyrene	X										X	X
Polycarbonate	X		X				X				X	X
Metalized Polycarbonate			X				X				X	X
Polypropylene	X		X								X	
Mica												
Dipped	X		X							X	X	
Molded	X		X								X	
Transmitting			X									
Glass												
Glass					X							
High-*K* Glass					X							
Ceramic												
Disc				X		X					X	
Tubular				X		X			X			
Monolithic				X		X			X		X	
Aluminum												
Twist Mount	X		X					X		X	X	
General Purpose	X		X				X	X		X	X	
Miniature	X		X					X		X	X	
Computer-Grade	X		X				X	X		X	X	
Tantalum												
Foil			X				X	X			X	
Wet Anode							X	X			X	
Dry Anode	X				X			X			X	X
Oil-Filled												
AC		X	X				X	X	X		X	
DC		X	X				X		X		X	
Energy Storage		X	X				X				X	

Variable Capacitors

Variable capacitors serve two circuit functions—tuning and trimming. Tuning capacitors are used to adjust, change, or vary the resonant frequency of an oscillatory or resonant L–C or R–C circuit, intermittently, or more or less continuously. There are two principal types: Air dielectric capacitors, and semiconductor voltage-variable capacitors called *varactors* or *tuning diodes*.

Trimming capacitors, or trimmers, are small-value adjustable capacitors used (usually in parallel) with larger-value tuning or fixed capacitors to make fine adjustment of the total capacitance in the circuit for calibration, tracking, and neutralizing. There are several types, classified by dielectric—air, mica, ceramic, glass, and plastic.

2.1 AIR DIELECTRIC VARIABLE AND TRIMMER CAPACITORS

Air has low insulation resistance compared to other dielectric materials, and the dielectric constant is about as low as dielectric constants go, but air is still useful as a dielectric material. Its stability is excellent, the power factor is about zero, and it is cheap.

Air dielectric variable capacitors are of two types: Fixed air-gap/variable-plate area, and fixed-plate/variable air-gap. Most are of the first type; the second construction is used for neutralizing capacitors.

Tuning capacitors (Fig. 2–1) are designed for frequent manual adjustment. Capacitance adjustment is made by varying the effective

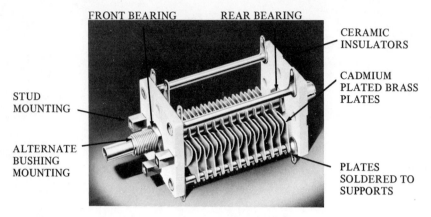

FRONT BEARING REAR BEARING

CERAMIC INSULATORS

CADMIUM PLATED BRASS PLATES

STUD MOUNTING

ALTERNATE BUSHING MOUNTING

PLATES SOLDERED TO SUPPORTS

Fig. 2-1 Single-section variable air capacitor. Series 23000 capacitor shown two-thirds size has maximum capacitance of 54.5 pF and 2000 v peak rating. *(Courtesy of James Millen Mfg. Co.)*

plate area. These capacitors consist of two sets of parallel intermeshed plates, one set fixed and the other set connected to a bearing-mounted shaft. Stamped cadmium plated steel is used for the frames in low-cost models; ceramic or porcelain end plate-and-metal post construction is used in better units. Regardless of construction, the frame must be rigid.

Rotor and stator plates are made of brass, copper, and aluminum, with aluminum being by far the most commonly used material. Brass and copper are plated with silver to increase surface conductivity, or plated with nickel or cadmium to prevent corrosion. The shape of the rotor plates can be contoured to obtain various capacitance versus rotation functions, such as straight-line frequency, linear capacitance, etc.

Plates are staked, soldered, or washer-spaced in capacitors with air gaps 0.015 inch to 0.265 inch. Some miniature trimmer capacitors with 0.010 inch air gaps are made with both rotors and stators machined out of solid, extruded, brass bar stock. (See Fig. 2-2.)

Tuning capacitors are made in a variety of styles, including single-section, multiple section, balanced-rotor dual section, and superheterodyne in which one of the two or three sections is cut—has less capacitance than the other section or sections—for tuning the local oscillator. Voltage ratings range from 600 vdc to 9000 vdc. High power symmetrically-configured air variable capacitors (Fig. 2-3) permit more compact component layout and better neutralizing; both one-to-one and geared drives are available. The range of standard varieties of air variable tuning capacitors is summarized in Table 2-1.

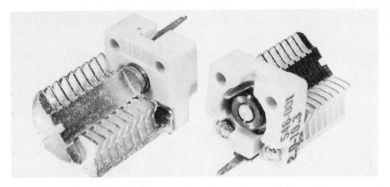

Fig. 2–2 Miniature precision air dielectric trimmer capacitors. Rotor and stator plate assemblies are machined from solid brass extrusions. Capacitors are shown enlarged two times. *(Courtesy of Erie Technological Products, Inc.)*

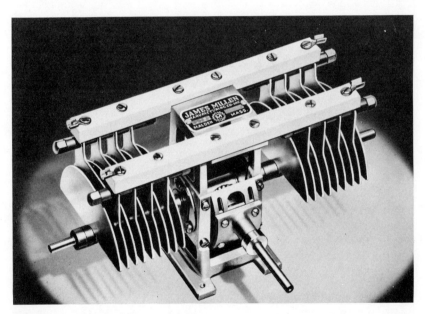

Fig. 2–3 Variable air transmitting capacitor. Millen 04000 series capacitor shown half size has 6000 V peak rating, and maximum capacitance of 48 pF per section. *(Courtesy of James Millen Mfg. Co.)*

Air trimmer capacitors differ from tuning capacitors in that they have a screwdriver slot for adjustment rather than an extended shaft and that, generally, they have lower capacitance.

Neutralizing capacitors are used in transmitting or receiving

TABLE 2-1 Air Tuning Capacitors†

Type	Plate Spacing	Working Voltage	Capacitance Range/Section		Plate Construction
			Typical Low-Value Capacitor	Typical High-Value Capacitor	
Transmitting	0.265 in.	9000 vdc	11 to 65 pF	40 to 130 pF	Polished alum., rounded edge
	0.171 in.	6000 vdc	13 to 52 pF	45 to 300 pF	"
	0.077 in.	3000 vdc	4.5 to 20 pF	16 to 150 pF	"
	0.066 in.	2250 vdc	4.5 to 20 pF	11 to 60 pF	Brass, cadmium-plated
General-Purpose	0.045 in.	1400 vdc	3 to 11 pF	7 to 39 pF	"
	0.022 in.	850 vdc	2 to 26 pF	14 to 339 pF	Brass, silver-plated
	0.020 in.	800 vdc	3 to 16 pF	10 to 145 pF	"
	0.015 in.	600 vdc	2 to 4.5 pF	5 to 110 pF	Brass, cadmium-plated

†Broadcast types are not included. Capacitors are made in single and balanced rotor dual section styles

circuits to feed a small portion of the stage output signal back to the input to prevent regeneration. Although any trimming capacitor can be used for neutralizing a receiver or low-power transmitter circuit, special neutralizing capacitors (Fig. 2–4) are used where high rf power and high dc voltages are present. There are two types: Disc and telescoping cylinder. The movable disc of the former is mounted on a fine pitch screw for close adjustment. A rotor lock nut is provided, as well as a positive mechanical stop to prevent a plate-to-plate short, which could be spectacular. In the telescoping cylinder type, an inverted cup-shaped electrode is adjusted up and down a threaded rod inside a larger cup.

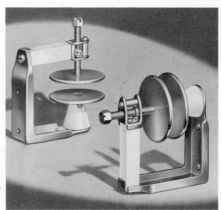

Fig. 2–4 Neutralizing capacitors, shown one-third size. *(Courtesy of James Millen Mfg. Co.)*

2.2 MICA TRIMMERS

Compression mica trimmers (Fig. 2–5) are inexpensive, they are nonlinear (usually unimportant), and they drift. They are also called *padders,* as they have been widely used in conjunction with radio receiver tuning capacitors to "pad" or trim the capacitance of the section of the air capacitor in order to obtain proper tracking in multiple-stage tuning. Capacitance values range from 1 to 3000 pF, but the range of any single trimmer is 10 to 1 for low values and down to 3.6 to 1 for high values.

A mica trimmer capacitor consists of a thin film, or leaf, of mica between two spring metal conduction plates, all mounted within a ceramic container (open at the top) or on a ceramic base. An insulated adjusting screw inserted through holes in the centers of both plates and the mica leaf, and threaded into a bushing in the ceramic base, provides variable compression on the formed metal

Fig. 2–5 Compression mica trimmer capacitors, shown full size. Maximum to minimum capacitance range is traversed with three turns of screws. Maximum capacitance for style capacitor ranges from 130 pF to 3055 pF depending on the number of plates. *(Courtesy of Arco Electronics)*

plates. Varying the plate separation changes the capacitance. The range of maximum to minimum capacitance is covered in 3 turns of the adjusting screw. For higher values of capacitance, plates and mica leaves are stacked, with plates alternately connected together as in fixed mica capacitors.

The base is a low-loss ceramic. Solder lugs are provided for connection. Mica trimming capacitors are mounted by two screws through holes in the ceramic base; clearance must usually be provided in the mounting surface to clear a threaded bushing and the lower end of the adjusting screw.

The dielectric is not mica alone, except when the plates are fully compressed at maximum capacitance; at other settings, the dielectric is air and mica in series, and the cross section of the air dielectric is actually wedge-shaped, which accounts for a good part of the nonlinearity of these trimmers.

PC board versions have terminals that can be snapped into board holes; however, the required board holes are rectangular in shape and not on standard grid centers. Mica trimmers are available in ganged multiple units, with mounting brackets, and with knurled-shaft handles for finger adjustment.

Hermetically sealed compression mica trimmers alleviate some of the problems inherent in all types of open adjustable capacitors.

A drawn-brass nickel-plated enclosure and an effective rotary shaft seal protect the plates and mica from moisture, atmospheric pressure variation, fungus, salt spray, corrosive atmospheres, and dust. The case also provides electrostatic shielding.

2.3 DISC CERAMIC TRIMMERS

Ceramic disc trimmers (Fig. 2–6) provide good stability and dependability under adverse operating conditions. Applications include computers, and communications, radar, aerospace, and test equipment. Subminiature types are used in electronic quartz watches. Capacitance of ceramic disc range from 1.5 to 1100 pF with a 5 to 1 range on a typical trimmer.

Fig. 2–6 Disc ceramic trimmers, shown full size. From left, Erie series 518, 538, 539, and 557. *(Courtesy of Erie Technological Products, Inc.)*

A ceramic disc trimmer contains two basic parts: Rotor and stator. (See Fig. 2–7.) A semicircular silver electrode is deposited, painted, or screened on the top surface of the rotor.

The rotor is a thin-lipped disc made of Class I or Class II ceramic dielectric material selected for the desired dielectric constant and thermal properties. A similar electrode is applied to the stator, which is also the shell of the capacitor. Both electrodes are bonded to the substrates.

As the rotor is turned, the effective capacitor plate area is changed. (The configuration is similar to that of a variable air capacitor.) Maximum to minimum capacitance change is produced in one-half turn of rotation. The shape of the stator electrode will be semicircular if maximum capacitance for the selected dielectric is desired; otherwise it can be considerably smaller.

For electrical stability, rotor and stator faces must be in intimate contact. A series intermittent air gap, for example, would have a disastrous effect on the capacitance rating; the steatite base and titanium dioxide rotor are lapped optically flat to provide minimum air

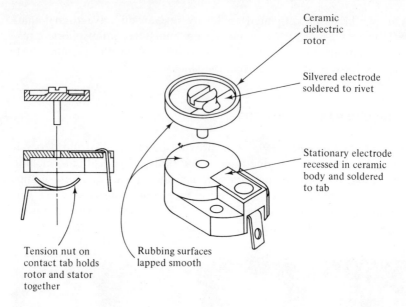

Ceramic
dielectric
rotor

Silvered electrode
soldered to rivet

Stationary electrode
recessed in ceramic
body and soldered
to tab

Tension nut on
contact tab holds
rotor and stator
together

Rubbing surfaces
lapped smooth

Fig. 2–7   Disc ceramic capacitor construction.

gap, and the rotor and stator are held together by spring pressure. However, too much spring pressure will cause galling and excessive wear and will require too much torque for precise adjustment. Spring pressure becomes a compromise between stability, operation life measured in rotations, and tuning ease. Terminals are silver-plated nonferrous metals.

Ceramic disc trimmers are made in standard, miniature, and subminiature sizes. They are available in larger capacitances than the tubular type and in a wide range of mechanical variations.

2.4 TUBULAR TRIMMER CAPACITORS

Tubular trimmer capacitors (Fig. 2–8) are used for fine adjustment of small capacitance values. Whereas a disc ceramic or air trimmer goes from maximum to minimum capacitance in one-half turn, a tubular trimmer covers the range in many turns of the adjustment screw. Tubular trimmers are available in several sizes and mounting arrangements, and they are made with different dielectric materials.

The capacitor consists of a glass, quartz, or ceramic tube on which a metal electrode is applied. Within this tube, moving in and out of the metalized silvered area, is a second electrode on a lead

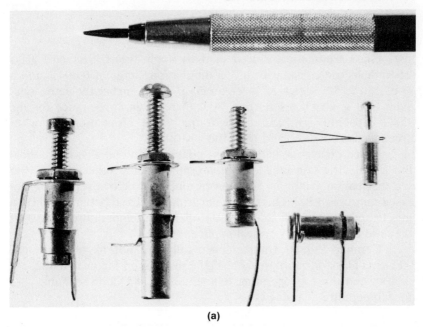

(a)

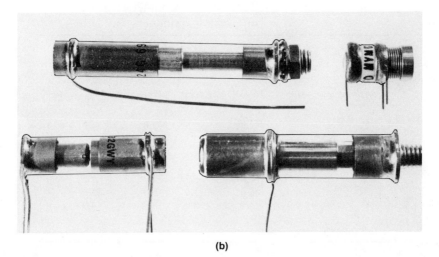

(b)

Fig. 2-8 Tubular trimmer capacitors. **(a)** Ceramic and teflon (far right) dielectric tubular capacitors. These capacitors utilize the adjustment screw as the center electrode. *(Courtesy of Centralab Electronics Div., Globe-Union, Inc.)* **(b)** Glass and ceramic (upper right) dielectric tubular capacitors. The center electrode of these capacitors is a machined piston attached to the lead screw. *(Courtesy of JFD Electronics Co.)* All capacitors shown full size.

screw. This electrode can be either a close-fitting piston or the lead screw itself.

Glass trimmers are used in most applications. Standard glass has a dielectric constant of 6.7, a dissipation factor of 0.0012, and a Q of 350 to 500. High-K, or green glass, has a higher dielectric constant of 8.3, which results in a 30% increase in capacitance for the same capacitor size and configuration. The dissipation factor of green glass is 0.0007, and Q is 500 to 1000.

Fused quartz (actually a glass containing no alkalis, only silica) has a dielectric constant lower than either of the other glasses, but the dissipation factor is also lower, only 3.8. However, the Q of fused quartz is 1500. Quartz tubular trimmers have better stability, a low temperature coefficient, and an operating temperature of 150°C; they are the choice for critical applications.

Ceramic tubular trimmers generally are not as temperature-stable as glass or quartz trimmers; the dissipation factor is 0.0025, Q is 400, and the top operating temperature is 85°C. (See Table 2–2 for temperature characteristics.)

TABLE 2-2   Temperature Characteristics
of Ceramic Trimmers†

	Percent Capacitance Change from Value @ 25°C					
	-55°C		*+85°C*		*+125°C*	
Dielectric Type Code	*Min.*	*Max.*	*Min.*	*Max.*	*Min.*	*Max.*
A	−4.5	+2.0	−2.5	+2.0	−4.2	+3.4
B	−1.0	+3.5	−2.5	−0.5	−4.2	−0.8
C	−1.0	+6.5	−4.0	−1.0	−6.7	−1.7
D	+1.5	+7.0	−5.0	−1.5	−8.5	−2.5
E	+1.75	+8.0	−5.75	−1.75	−9.5	−2.8
F	+6.0	+16.0	−11.0	−6.0	−15.0	−9.0
G	0.0	+14.0	−8.0	−3.0	−14.0	−5.0

†(Erie Technological Products, Inc.)

Polytetrafluoroethylene (Teflon ®) is used as a dielectric for subminiature trimmers. The material provides the highest Q, 2000, high dielectric strength and insulation resistance. Polystyrene is used in low-cost tubular trimmers for some consumer product applications.

Tubular trimmers are normally mounted perpendicularly to the mounting surface by means of a tension locknut and adjustment screw.

Methods of manufacture vary. In the simplest type (Fig. 2–9), the outer electrode is formed by firing a layer of metallic silver on a glass tube. This silvered area is then solder-coated to retard oxidation. The adjustment screw itself is the inner electrode.

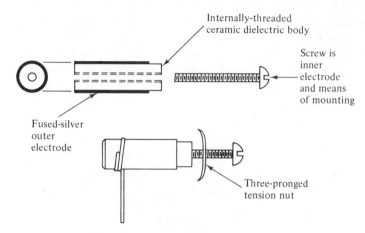

Fig. 2–9 Ceramic tubular trimmer capacitor construction.

Maximum capacitance occurs with the screw all the way in; specified minimum capacitance is reached with three threads meshed.

The design of piston tubular trimmers is more complex. Where extension of the screw above the panel in standard panel mounting can be tolerated, the economical design shown in Fig. 2–10 is used. The threads of the adjustment screw engage mating threads in both the mounting bushing and an axially loaded spring tension nut captive in the mounting base.

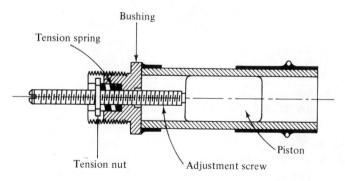

Fig. 2–10 Standard panel mount piston capacitor construction. *(Courtesy of JFD Electronics Co.)*

Telescoping adjustment mechanisms are used where protruding and adjustment screws cannot be allowed (Fig. 2–11). The threads of the screw engage mating threads of a slotted flange extension of the inner bushing. A tensioning ring in a slot on the flange imposes a desired torque on the threads. In the design of Fig. 2–12, torque is imposed on the screw by two axially spring-loaded tension nuts riding along flanges in 90 degree opposed channels of the mounting bushing. These trimmers are made in open and sealed versions; in the sealed styles, an O-ring riding in a groove in the adjusting screw head forms an effective seal.

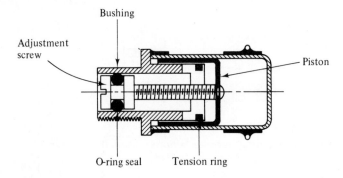

Fig. 2–11   Miniature telescoping piston trimmer construction. *(Courtesy of JFD Electronics Co.)*

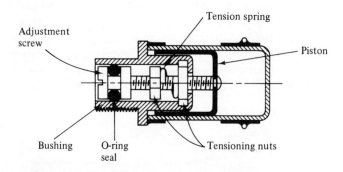

Fig. 2–12   Miniature 500 cycle life telescoping piston trimmer construction. *(Courtesy of JFD Electronics Co.)*

None of the preceding designs have a life of more than 500 adjusting cycles. The sliding travel adjustment mechanism shown in Fig. 2–13 has a life of 10,000 cycles. Two tension nuts, one of which is an integral part of the inner piston assembly, ride in 180 de-

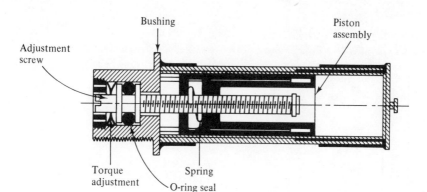

Fig. 2–13 Sliding-travel 10,000 cycle life piston trimmer construction. *(Courtesy of JFD Electronics Co.)*

gree opposed slots in an elongated flange extension of the mounting base.

A spring between the nuts prevents backlash with the adjustment screw, which remains in a fixed position in the adjustment well. Torque is controlled at the end of the bushing and can be preset to any value between 1 and 10 inch-ounces. Rotation of the adjustment screw actuates the linear travel of the piston within the dielectric.

Electrodes are applied in a variety of patterns. Most common is the *fully metalized* pattern which produces a straight-line capacitance change with piston travel. A *closed-end* metalized sealcap construction produces a similar capacitance characteristic, with twice the working voltage, and can be encapsulated or potted. The tubular trimmer can function as two capacitors with *split-stator* metalizing and a grounded piston. Tracking is good, and the metalized segments need not be the same length.

Table 2–3 provides a comparison of tubular trimmer capacitor types.

2.5 VARACTORS

Varactors, voltage variable capacitors, tuning diodes, or silicon diode capacitors (Fig. 2–14) are two-terminal semiconductor devices that put the junction capacitance of a reverse-biased PN junction to work. The capacitance of these devices varies inversely with the applied reverse bias voltage. A 2 to 1 capacitance change for specified

TABLE 2-3 Characteristics and Typical Values of Trimmer Capacitors†

Type and Dielectric	Capacitance, Typical Low-Value Trimmer	Range, Typical High-Value Trimmer	Working Voltage (Max.)	Operating Temperature (Max.)	Temperature Coefficient	Q, Minimum
Air Trimmer, Plate	1.2 to 3.5 pF	4 to 35 pF	325 vdc	+125°C	0 ± 80 ppm/°C	1500
Neutralizing, Disc	3.5 to 18 pF	—	(1)	—	0 ± 80 ppm/°C	1500
Neutralizing, Cylinder	3.5 to 16 pF	—	4000 v peak	—	0 ± 80 ppm/°C	1500
Mica Compression, Open	1 to 12 pF	390 to 1400 pF	175 vdc 250 vdc 500 vdc	+85°C	—	—
Mica Compression, Hermetically Sealed	1 to 12 pF	1400 to 3000 pF	250 vdc 500 vdc	+85°C	—	—
Ceramic Disc, Standard	1.5 to 7 pF	11 to 110 pF	250 vdc 500 vdc	+125°C	(3)	500
Ceramic Disc, Miniature	1.5 to 8 pF	15 to 60 pF	100 vdc 200 vdc 350 vdc	+125°C	(3)	500
Ceramic Disc, Sub-	1 to 3 pF	7 to 40 pF	25 vdc		(3)	(2)

TABLE 2-3 Continued

Type and Dielectric	Capacitance, Range: Typical Low-Value Trimmer	Typical High-Value Trimmer	Working Voltage (Max.)	Operating Temperature (Max.)	Temperature Coefficient	Q, Minimum
Tubular, Air miniature	0.8 to 10 pF	1 to 20 pF	250 vdc, 50 vdc, 100 vdc	+125°C	0 ± 20 ppm/°C	3000
" Glass	0.8 to 4.5 pF	1 to 30 pF	750 vdc	+125°C	0 ± 50 ppm/°C, 0 ± 100 ppm/°C	350 to 500
" Quartz	0.6 to 1.8 pF	0.8 to 16 pF	750 vdc	+150°C	25 ± 25 ppm/°C	1500
" High-K Glass	0.8 to 5.5 pF	1 to 38 pF	750 vdc	+125°C	0 ± 50 ppm/°C, 0 ± 100 ppm/°C	500 to 1000
" Sealed Glass	1 to 14 pF	1 to 90 pF	750 vdc, 1250 vdc	+125°C	0 ± 50 ppm/°C, 0 ± 100 ppm/°C	350 to 1000
" Ceramic	0.5 to 6 pF	3.5 to 20 pF	500 vdc	+85°C	P120, N220, N400	400
" Teflon	0.25 to 1.5 pF	–	600 vdc, 350 vdc, 500 vdc	+125°C	–	2000
" Polystyrene	0.5 to 5 pF	1 to 8 pF		+85 °C	–	1000

†*Notes:* 1. Minimum air gap 0.4 inches.
2. 1350 to 100 in frequency range 1 to 100 MHz.
3. See Table 2-2.

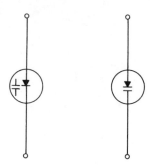

Fig. 2-14   Varactor symbols.

voltage change is typical, and they are available in most of the same capacitance values as air-dielectric variable capacitors (Table 2-4). Tuning diodes go under many trade names, Epicap (Motorola), Varicap (TRW), etc.

But, you don't get something for nothing: Varactor capacitance is temperature-sensitive and requires compensation. And the voltage supply must be stable — *very* stable.

Definitions.

TUNING RATIO (TR): The tuning, or capacitance, ratio is the ratio of capacitance obtained with two values of applied bias voltage.

FREQUENCY RATIO: The frequency ratio is equal to the square root of the tuning ratio (no stray circuit capacitance is assumed).

QUALITY FACTOR (Q): The Q of a varactor (or any capacitor) is the ratio of a capacitive reactance to the series resistance.

Air variable tuning capacitors have a Q of 1000 or so. The Q of circuits using these capacitors is mainly determined by coil Q, but in a varactor-tuned circuit, the much lower Q of the varactor sets up a different problem. The Q of the varactor is not constant but depends on both bias voltage and frequency (Fig. 2-15) and temperature. The Q falls off at high frequencies because of the series bulk resistance of the silicon used in the diodes, and it also falls off at low frequencies because of the back resistance of the reverse-biased diode; Q varies directly with bias voltage and inversely with temperature. A simplified circuit of a varactor is given in Fig. 2-16; lead inductance and case capacitance is neglected.

The temperature constant (TC) of a varactor is a function of applied bias. A forward-biased diode provides a simple method of

TABLE 2-4 Typical Range of Characteristics of Varactor Diodes

	Device Series				
	1N5139	1N4801	1N950	1N5681	MV1400 (MOT) PQ (TRW)
Capacitance					
−4 vdc @ 1 MHz	6.8 to 47 pF	6.8 to 100 pF	35 to 100 pF	6.8 to 100 pF	—
−8 vdc @ 1 MHz	—	—	—	—	120 to 550 pF
Capacitance Range					
−0.1 vdc to max. working voltage, typical	—	—	6 to 88 pF 22 to 120 pF 46 to 240 pF	—	—
Tuning Ratio Min.	2.7 to 3.4	2.2 to 2.4	—	3.1 to 3.2	2.5 to 3.0
Q Min.	200 to 350 @ 50 MHz	15 @ 50 MHz 750 @ 1 MHz	175 to 3360 @ 5 MHz	150 to 600 @ 50 MHz	100 to 200 range 10 to 25 MHz
Max. Working Voltage	60 vdc	20 to 100 vdc	25 to 130 vdc	40 to 60 vdc	60 to 90 vdc
Power Dissipation, Max.	400 mw	600 mw	400/600 mw	400 mw	5 w
Operating Temperature	−65°C to +150°C	−65°C to +150°C	−65°C to +150°C	−65°C to +150°C	−65°C to +150°C
Case Capacitance, Typical	0.25 pF	0.25 pF	0.25 pF	0.25 pF	—
Lead Inductance, Typical	5 nH	5 nH	5 nH	5 nH	—
Cost, each	$4.65	$3.60	$3.40	$6.55	$8.95

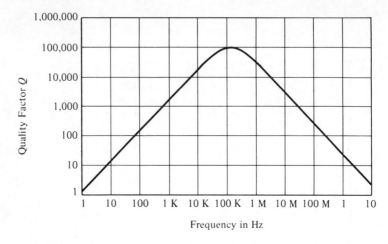

Fig. 2–15 Typical varactor *Q* versus frequency characteristic (1N5139 Epicap tuning diode). *(Courtesy of Motorola Semiconductor Group.)*

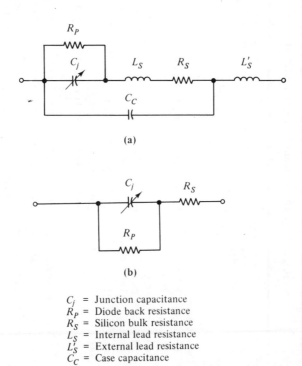

C_j = Junction capacitance
R_P = Diode back resistance
R_S = Silicon bulk resistance
L_S = Internal lead resistance
L_S' = External lead resistance
C_C = Case capacitance

Fig. 2–16 (a) Varactor equivalent circuit; **(b)** Simplified equivalent circuit neglecting case capacitance and lead inductance. *(Courtesy Motorola Semiconductor Group.)*

temperature compensation. In the circuit of Fig. 2–17, the voltage drop of the forward-biased diode decreases as the temperature rises, thus applying a changing voltage to the varactor.

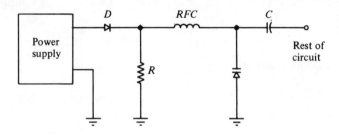

Fig. 2–17 Simple varactor temperature compensating circuit. *(Courtesy of Motorola Semiconductor Group.)*

The source of the capacitance-determining bias voltage must be extremely stable. Temperature stability is critical. Varactor tuning stability can be only as good as the power supply.

The voltage supply can be regulated several ways. Zener diodes are relatively inexpensive but have a limited voltage range. Table 2–5 compares a variety of regulators.

TABLE 2-5   Comparison of Power Regulators
(Courtesy of Motorola Semiconductor Group)

Device	Voltage Range	Temperature Range	Voltage ppm/°C Maximum TC	Voltage ppm/°C Typical TC	Relative Cost
1N5260 Zener	33	− 65 +200°C	975	975	Low
1N4752 Zener	33	− 65 +200°C	850	850	Low
1N3157 Temperature Compensated Zener	8.4	− 50 +125°C	10	10	High
MC1723 Regulator	37	− 55 +125°C	20	12	Medium
MFC6030 Functional Regulator	32	0° +70°C	50	15	Low
MC7800 Fixed Voltage Regulators	28	0° +125°C		40–60	Medium
MVS460 TO–92 Regulator	31 V	0° +70°C	100 +50	25	Low

The variable tuning resistor is less critical, as it performs merely as a voltage divider. Wire wound, and high-quality cermet film variable resistors are suitable. A linear potentiometer should have a TC of ±150ppm/°C; a taper potentiometer, a TC of $+50$ppm/°C or better.

2.6 WHO MAKES WHAT

Table 2–6 lists some of the major variable capacitor manufacturers.

SUGGESTED READINGS

Dummer, G. W. A., and Harold M. Nordenberg, *Fixed and Variable Capacitors*. McGraw-Hill Book Co., New York, N. Y., 1960, pp. 201–57.

Johnson, Doug, and Roy Hejhall, *Tuning Diode Design Techniques*. Motorola Semiconductor Products, Inc., Phoenix, Ariz., 1973.

Motorola Semiconductor Products, Inc., *Epicap Tuning Diode Theory and Application*. Phoenix, Ariz., 1971.

CATALOGS

Arco Electronics, Great Neck, N. Y.

Centralab Electronics Division, Globe Union, Inc., Milwaukee, Wisc.

Erie Technological Products, Inc., Erie, Pa.

James Millen Manufacturing Co., Inc., Malden, Mass.

JFD Electronics Corp., Brooklyn, N. Y.

TABLE 2-6   Variable Capacitor Manufacturers

	Arco Electronics	Centralab	Erie	International Rectifier	JFD	James Millen	J. W. Miller	Motorola Semiconductor	Sprague	Texas Instruments	TRW Semiconductor
Air Tuning											
General Purpose						X	X				
Transmitting						X					
Air Trimmer											
Plate			X			X					
Disc Neutralizing						X					
Cylinder Neutralizing						X					
Mica Compression Trimmers											
Open	X						X				
Hermetically Sealed	X										
Ceramic Disc Trimmers											
Standard			X		X						
Miniature			X		X						
Subminiature			X		X						
Tubular Trimmers											
Air					X						
Glass			X		X				X		
Quartz			X		X				X		
High-K Glass					X				X		
Glass, Sealed					X						
Ceramic		X			X						
Teflon			X								
Polystyrene			X								
Varactor Diodes											
All Types				X				X		X	X

CHAPTER THREE

Transformers and Inductors

In its simplest form, a transformer is a two winding device; the two windings are inductively coupled by mutual inductance in an air-cored transformer or by a common magnetic core in iron or ferrite-cored transformers. An ac voltage applied to one winding will cause an ac voltage of the same frequency to be induced in the other winding. The magnitude of the voltage induced in the second winding is approximately equal to the applied voltage times the ratio of the number of turns on the two windings — approximately, but reduced slightly by losses in the transformer.

Transformers are made in great variety to meet application requirements. Those shown in Fig. 3-1 are typical of the types encountered in low power electronic applications. The trend in transformer design is to miniaturization (Fig. 3-2).

Power transformers change (transform) the ac supply voltage to the amplitude or amplitudes required by the load circuits. Audio and communications transformers transmit information or intelligence signals of varying frequency, amplitude, and waveshape, between pieces of equipment or between circuits in a single piece of equipment.

An inductor consists of a two-terminal single winding wound on an air or magnetic core, depending on the inductance and operating characteristics desired. Inductors serve a variety of purposes: Power supply filters, oscillators, tuned circuits, and frequency discriminating filters.

Transformer Operation. An ideal transformer has infinite winding reactances and zero leakage inductance, core loss, winding

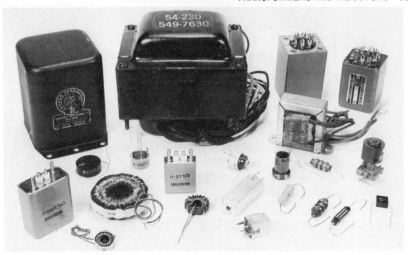

Fig. 3-1   Transformers and inductors.

Fig. 3-2   Ultra-miniature audio and pulse transformers, shown approximately full size. *(Courtesy of Pico Electronics, Inc.)*

capacitance, and winding resistance. Voltage ratios equal turns ratios under all loading conditions, and the current ratios in any two windings are inversely proportional to the turns ratios. Such perfection is not achieved, but modern iron-core transformers come close, and analysis on the basis of an ideal transformer gives useful approximations for practical design and specification purposes.

Figure 3-3 shows an ideal two-winding transformer. Such a transformer provides isolation between the primary and secondary winding in addition to an N_1/N_2 voltage ratio. An N_1/N_2 can also be obtained with an autotransformer as shown in Fig. 3-4. In two-

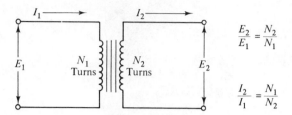

Fig. 3–3 Ideal two-winding transformer.

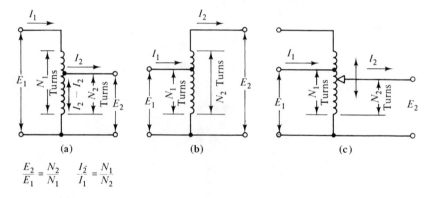

$$\frac{E_2}{E_1} = \frac{N_2}{N_1} \qquad \frac{I_2}{I_1} = \frac{N_1}{N_2}$$

Fig. 3–4 Autotransformer. **(a)** Tapped primary, **(b)** tapped secondary, **(c)** variable autotransformer.

winding transformers and autotransformers, the primary and secondary currents are 180° out of phase; in the autotransformer, the current flow in the portion of the winding common to the primary and the secondary is the difference between the two.

Most transformers used with electronic equipment have separate windings, because the circuit isolation provided is at least as important as changing voltage levels. Autotransformers are used primarily as supply-voltage adjusting devices.

Practical transformers are something else. There are winding resistances and inductances, leakage inductances, capacitances, and core loss, as shown in Fig. 3–5. If you want to design transformers, you have to start here, but for practical procurement, you can work most of the time with ideal transformer parameters.

If an ideal transformer has its secondary winding loaded with a resistance as shown in Fig. 3–6, an equivalent load resistance is reflected to the primary winding. If there were no secondary load,

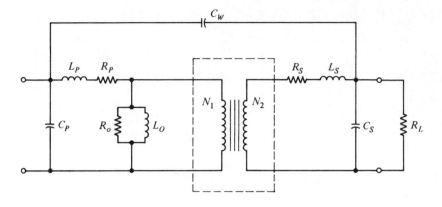

L_p = Equivalent primary leakage inductance
R_p = Resistance of the primary winding
R_S = Resistance of the secondary winding
L_S = Equivalent secondary leakage inductance
R_O = Equivalent core loss resistance
L_O = Inductance of the primary winding
C_p = Primary equivalent lumped capacitance
C_S = Secondary equivalent lumped capacitance
C_W = Equivalent lumped interwinding capacitance
R_L = External load resistance on secondary winding

Fig. 3-5 Simplified transformer equivalent circuit.

there would be no current flow in the primary (in practical transformers there is a no-load primary current because of the finite winding resistance, etc.). A transformer reflects (transforms) a load placed on its secondary. The turns ratio between the windings (not the number of windings on either) determines the load impedance. From a practical standpoint, there must be enough primary turns to keep primary reactance high compared to the reflected load impedance.

The load, or impedance ratio, is equal to the square of the turns ratio. For transformer versatility, either or both primary and secondary windings may be supplied with taps, allowing one transformer to satisfy many applications.

Core Materials. The usefulness of iron as a magnetic core material is limited, because iron is also a good conductor of electricity. The current flowing in the primary winding induces a current flow in the core just as it does in the secondary winding. If the

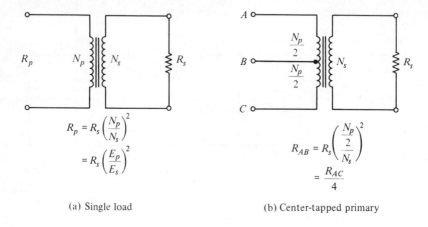

(a) Single load

(b) Center-tapped primary

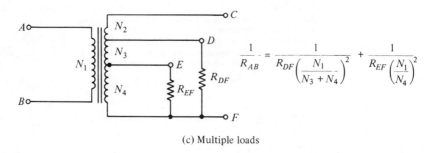

(c) Multiple loads

Fig. 3–6 Loading an ideal transformer.

transformer core were a solid block of iron, it would behave as a single shorted turn. This induced current flow in the magnetic core of a transformer is called an *eddy current,* and it creates a magnetic field of its own that opposes the field induced by the primary winding current.

To reduce eddy currents at power and audio frequencies, transformer cores are made of thin sheets, or laminations, of steel stacked together but separated by insulating glue. This breaks up the electrical continuity of the core. The magnitude of the induced eddy currents varies directly with the frequency of the primary current, and laminating the transformer core works reasonably well up to 100 kHz.

Above that frequency, the eddy currents within each lamination grow so large that the primary field is all but cancelled. Powdered metal cores—grains of ferrous metal compressed and held together by an insulating binder—are useful as transformer cores into the MHz range, but the material also loses magnetic properties by its structure. For many high-frequency applications, ferrites—iron oxide materials that combine substantial magnetic strength with extremely high inherent electrical resistance—are used for transformer and inductor cores. For many high-frequency transformers and inductors, the powdered or ferrite core is in the shape of a cylinder inside the coiled winding and does not form a closed magnetic path.

Ferrites. Ferrites are artificial nonmetallic magnets. They are composed of a compressed and sintered powder consisting chiefly of an iron oxide Fe_3O_4 with other metallic ions present to control the magnetic properties. Ferrites are hard, dense, and brittle ceramics and are almost completely crystalline. Some small ferrites are cut from larger pieces of single-crystal material.

Soft ferrites are used in transformer cores. Their high resistivity keeps eddy–current losses low at high frequencies, which reduces heating. These ferrite cores make possible miniaturized transformers for many high-frequency applications including television deflection yokes, radio antennas, pulse transformers, inductances, radio–frequency transformers, and tape recorder record and playback heads. Soft ferrites are made from combinations of nickel, zinc and iron oxides; manganese, zinc and iron oxides; and nickel, copper, zinc and iron oxides.

Square-loop ferrites are used as computer memory cores and other switching and control applications. These ferrites have two stable states of magnetization. They can be switched from being magnetized in one direction to the other direction, depending on the polarity of the current in the magnetizing wire, and the ferrites will remain magnetized in either of the directions after the magnetizing current is removed, until magnetized in the other direction by a current of the opposite polarity. Meanwhile, the polarity of the magnetization, representing a 1 or 0, can be read out.

These ferrites are in the shape of small toroids, so small that in some sizes there would be a million of them in a pound. These ferrites are made of manganese and iron oxides.

Ferrites have also been developed for switching and control at microwave frequencies. These ferrites, composed of aluminum, nickel, zinc and iron oxides, or from magnesium, manganese and iron oxides, have low electrical losses at microwave frequencies.

Some ferrites are permanent magnets—ceramic magnets—and they are widely used in speakers and small dc motors in which a ferrite magnet replaces a copper winding. Applications for these motors include portable tools, and automotive accessories.

3.1 DEFINITIONS

AUTOTRANSFORMER: A transformer with a single electrical winding that acts as both primary and secondary. The whole winding may be the primary, and a tapped part of the winding, the secondary for a step-down voltage ratio, or vice versa for step-up. The tap is usually adjustable.

CHOKE: An inductor used to impede the flow of ac or pulsating dc.

CHOKE, SWINGING: An iron-cored inductor, usually containing an air-gap in the magnetic core circuit; used in rectifier filter circuits. Effective inductance of a swinging choke varies with the current flowing through it.

CORE LOSS: The energy lost in a magnetic core due to eddy currents circulating in the core. The energy is dissipated as heat.

DECIBEL (dB): The unit for expressing differences in power levels and transmission gain or loss. Decibels equal $10 \log_{10} P_1/P_c$. Decibels are often measured relative to a standard reference level, such as one milliwatt. In that particular case, the term used is dBm.

EDDY CURRENTS: Currents induced in a magnetic structure, such as a transformer core.

FERRITE: An artificial nonmetallic magnetic material composed of compressed and sintered iron oxide with other metal ions to control magnetic properties. Unlike ordinary magnetic materials, ferrites have electrical resistivity.

FILAMENT TRANSFORMER: A transformer used to supply power to the filaments of vacuum tubes. They are also used for other low-voltage power applications.

FILTER CHOKE: An iron-core inductor used in a rectifier filter circuit to impede the passage of the alternating component of the rectified current.

HARMONIC DISTORTION: The alteration of a waveform by the addition of waveforms of harmonic frequency, usually as a result of nonlinearities in the characteristics of the component, circuit, or equipment.

HUM: An inductively or electrostatically picked-up extraneous signal, usually from ac power sources or rectifiers.

HYSTERESIS: Time lag between the magnetization of a ferrous material . and the magnetizing force, caused by molecular friction.

INDUCTOR: Any coiled conductor, with or without a magnetic core, that is used to introduce inductive reactance into a circuit.

INPUT TRANSFORMER: Any transformer that connects a transducer, transmission line, or any other piece of equipment to the input circuit of a piece of equipment.

INTERSTAGE TRANSFORMER: Any transformer that is used to transfer the output signal of one circuit to the input of another circuit within a piece of equipment.

LINE TRANSFORMER: Any transformer used to connect a signal circuit of a piece of equipment to a transmission line, or used to interconnect transmission lines.

OUTPUT TRANSFORMER: Any transformer used to connect the signal of an output stage of an amplifier to a load.

PERMEABILITY: The measure of how much better a material is than air as a path for magnetic lines of force. Air has a permeability of one.

PLATE TRANSFORMER: A power transformer used to supply high-voltage power to the plate circuit of vacuum tubes.

POWER TRANSFORMER: Any transformer used to provide power at required voltage and current levels for the operation of electronic equipment.

PULSE TRANSFORMER: A wide-band transformer designed to transmit pulses with good waveshape fidelity, as compared to audio transformers which are designed to transmit continuous, essentially sinusoidal waveforms.

SATURATION, MAGNETIC: State of magnetism in a magnetic material beyond which it is incapable of increased magnetization.

3.2 POWER TRANSFORMERS

A power transformer is designed to operate most efficiently at a single frequency; e.g., 60 Hz, 400 Hz, or 2000 Hz; or over a band of power frequencies when operated from a variable-frequency power source. The design operating frequency has great effect on the size and weight of the transformer.

While capacitors and resistors are off-the-shelf components, power transformers are more likely to be custom units except in one-of-a-kind and low-quantity equipment designs. Iron-core transformers are bulky and heavy; specifying exactly what you need can save space and weight, a particularly rewarding activity in the design of portable and airborne equipment. There is great flexibility in the design of all iron-core transformers, and the possible permutations obtainable among their parameters is tremendous.

Because power transformers used with electronic equipment seldom operate into resistive loads, ratings are given in volt–amperes (VA). The VA of the secondaries of transformers connected to rectifiers with choke input or capacitor input filters can be larger than the primary VA.

Other factors affect the size of a power transformer besides operating and frequency and VA rating. For the same VA rating, a high-voltage transformer will be bigger because of the need for thicker insulation. Losses in a practical transformer generate heat; high ambient operating temperature requirements reduce allowable internal temperature rise, necessitating use of larger wire sizes, which produces a bigger transformer. If you require transformer efficiency approaching 100 percent, think big; normal-sized transformers have efficiencies of 80 to 90 percent, miniature power transformer efficiencies can be as low as 60 percent. Long life and tight regulation requirements also mean bigger transformers. Sealed, steel-cased transformers can weigh two to three times an equivalent open transformer. Size and weight can be reduced with the use of more efficient core materials (more expensive) that have low losses and high saturation density.

3.3 AUDIO TRANSFORMERS

These transformers are designed for audio, carrier, and broadband applications in the 20 Hz to 500 kHz frequency range. They are used to connect transducers to amplifiers, connect circuit stages, change voltage levels and match impedances, invert signal polarities, and provide electrical isolation. Audio transformers are designed for specific applications based on circuit function, signal level, source and load impedances, and frequency band. Parameters are usually selected by calculation (yours or the vendor's) and ultimate suitability determined by breadboarding.

Power Rating. The operating level restricts the choice of core material and determines the minimum physical size of the transformer. Operating level is expressed in decibels (dB) above or below a reference level of 1 milliwatt. For the same operating level, different core materials also result in wide variations in size.

Frequency Response. Of major importance in selecting an audio transformer is the required frequency response and the allowable waveform distortion. Very wide frequency response is desired in high fidelity transformers, but a narrower frequency range is mandatory in communications equipment where emphasis is not on fidelity but on intelligibility.

In low-level transformers (those operating at signal levels below 0 dB, or 1 milliwatt), the problem is to obtain good transmission at the low end. This requires high inductance as the gain falls off due to decrease in primary reactance. The response will be down 3 dB from the mid-range response at the frequency where $2\pi L = R$; L equals primary inductance and R equals source resistance plus the resistance of the primary winding. This frequency (f_1) is called the *low half-power frequency.*

At the high-frequency end, leakage reactance, shunt capacitance, and source and primary winding resistance form a low-Q series resonant circuit. Above the resonant frequency, response falls off rapidly. The frequency (f_2) at which the response is down 3 dB is called the *upper half-power frequency.* High-frequency response can be improved by reducing the number of turns, thus reducing L, but that then disrupts the low-frequency response.

Harmonic distortion is not usually a problem in low-level transformers because of low core saturation, but hum is a problem. To keep hum at a reasonably low level in relation to the signal (-60 dB), high permeability shielding is required. More than one layer of shielding material may be required, as well as careful location and orientation of the transformer in the equipment.

In output and other high-level transformers, high inductance is also required for good low-frequency response, but the problem is made more difficult by the flow of dc in the winding, which reduces the permeability of the core material. The solution is more core material, which is why big is usually better in output transformers. High end response is limited as for low-level transformers.

3.4 IRON-CORE INDUCTORS

Iron-core inductors or *chokes* (Fig. 3–7), as they are commonly called, usually consist of a single two-terminal winding on a magnetic core. Some chokes are made with split windings that may be connected in series, parallel, or series with a center tap, such as might be used in filters, tuned circuits, or coupling inductors. Designs vary, depending on the application, frequency, or frequency range, and depending on whether ac or both ac and dc must be carried. (See Fig. 3–8.) Iron cores, having permeability higher than air, provide higher values of inductance and Q. Adjustable inductance can be obtained by changing the width of an air gap in the core. Magnetic shielding and hum-bucking construction is used on better units. The level of dc flowing in the coil affects permeability and saturation.

Inductance is determined by the flux density in the core, which is a function of the ac current. Maximum flux density is limited by both allowable temperature rise and core material saturation. As

Fig. 3-7 Small inductors: (top) solenoid wound, (below) toroidal wound. *(Courtesy of J. W. Miller Div., of Bell Industries.)*

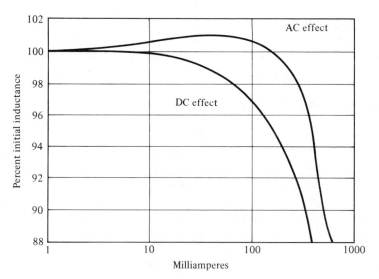

Change in initial inductance with AC and DC
excitation: microtran QGL series toroidal inductors

Fig. 3-8 Inductor change in initial inductance with ac and dc excitation for Microtran QGL series toroidal inductors. *(Courtesy of Microtran Co., Inc.)*

there will be a decrease in both inductance and Q with increasing dc current, both parameters are usually measured with no dc, except in the case of power supply filter inductors for which inductance is normally specified at some dc current level.

Q is determined by the interwinding capacitance and the effective resistance of the inductor, which is composed of winding resistance, and hysteresis and eddy current core losses.

Inductor design is essentially a function of operating frequency. Magnetic cores are used at audio and low carrier frequencies to combine large inductance and low dc resistance with reasonable size. Where "brute force" suppression of ac is the only requirement, laminated cores are used, resulting in the smallest size for the inductance. But the inductance value is not precise, Q is low, and any dc present in the winding will substantially reduce the inductance. If accurate inductance is required, permalloy powdered cores or ferrite cores must be used.

At high carrier and radio frequencies, air-core and slug-tuned inductors are used. (See Fig. 3–9.) Table 3–1 compares typical small radio-frequency inductors.

Fig. 3–9 Small solenoid wound rf inductors, shown full size. Cores are phenolic, powdered iron, ferrite, and air and phenolic. *(Courtesy of J. W. Miller Div., of Bell Industries.)*

Toroids. The coils of toroidal inductors (Fig. 3–10) are wound on doughnut-shaped cores. The cores are usually made of compressed molybdenum permalloy powder. High values of Q are obtained (Fig. 3–11) by using the largest wire-size possible while still getting the number of turns through the hole in the core needed for the required inductance.

TABLE 3-1 Radio–Frequency Inductors†

Series	Description	Size (in.)
T1, (JW Miller)	Powdered-iron toroidal core, subminiature encapsulated case	0.2 x 0.2 x 0.1
T2, (JW Miller)	Powdered-iron toroidal core, subminiature encapsulated case	0.2 x 0.2 x 0.1
T3, (JW Miller)	Powdered-iron toroidal core, subminiature encapsulated case	0.27 x 0.27 x 0.15
T4, (JW Miller)	Powdered-iron toroidal core, subminiature encapsulated case	0.38 x 0.38 x 0.19
Super Wee-Wee-Ductor (Nytronics)	Phenolic core, shielded and encapsulated, axial lead	0.14 d x 0.34 h
Super Wee-Wee-Ductor (Nytronics)	Powdered iron core, shielded and encapsulated, axial lead	0.14 d x 0.34 h
Super Wee-Wee-Ductor (Nytronics)	Ferrite core, shielded and encapsulated, axial lead	0.14 d x 0.34 h
RF Choke (JW Miller)	Powdered iron core, coils impregnated with moisture-resistant lacquer	varies; 0.16 d x $\frac{1}{4}$ l to 0.4 d x $\frac{3}{8}$ l

† Varieties described are only a small sampling of what is available.

Toroidal inductors are used in wave filters, oscillators, discriminators, and power supply filters in the 200 Hz to 200 kHz frequency range. Inductance is essentially independent of frequency, temperature, and ac and dc signal levels. Hum pickup is extremely low because of the symmetrical winding, and toroids can be stacked close together without the mutual inductive coupling of other inductor construction.

The inductors are available dipped, epoxy-molded, or hermetically sealed. Magnetic shielding is seldom required. Toroidal inductors can be obtained with taps; toroidal transformers are also made. Typical laminated and toroidal core inductors are compared in Table 3–2.

TABLE 3-1 *Continued*

Apparent Inductance	Rated Current mA dc	DC Resistance (Maximum)	Min. Q and Test Frequency	Cost, Each, Small Quantity
0.01 to 1.0 μH	3000 to 700	0.02 to 0.4 Ω	60 to 80 @ 50–150 MHz	$3–$9
0.1 to 10 μH	2200 to 280	0.04 to 2.6 Ω	55 to 70 @ 7.9 or 25 MHz	$7–$13
10 to 1000 μH	550 to 110	1.1 to 27 Ω	50 to 75 @ 7.9 or 25 MHz	$10
100 to 10,000 μH	280 to 55	6 to 180 Ω	50 to 80 @ 250 or 790 kHz	$12
0.1 to 0.82 μH	1720 to 720	0.11 to 0.64 Ω	36 to 42 @ 25 MHz	$2
1.0 to 12 μH	1260 to 300	0.21 to 3.7 Ω	36 to 51 @ 2.5 or 7.9 MHz	$2
15 to 10,000 μH	620 to 231	0.86 to 37 Ω	37 to 57 @ 0.25 to 25 MHz	$2
0.1 μH to 100 mH	3900 to 29	0.013 to 278 Ω	30 to 50 @ 79 kHz to 25 MHz	$1–$2

3.5 PULSE TRANSFORMERS

A pulse transformer is a wide-band transformer designed to transmit pulses with good waveshape fidelity rather than continuous sine waves. Features include high-voltage insulation between windings and to ground, low interwinding capacitance, and low winding reactance. Pulse transformers are normally used either of two ways: For coupling and in pulse-generating circuits.

Pulse Waveform. For a coupling application, source and load impedances, voltage levels, repetition rate, and the waveform of the output pulse must be specified, the latter correctly specified in pulse parameters (Fig. 3–12).

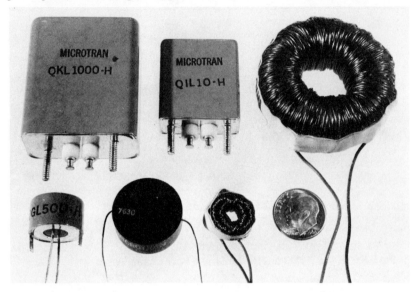

Fig. 3–10 Toroidal inductors, shown approximately full-size. *(Courtesy of Microtran Co., Inc.)*

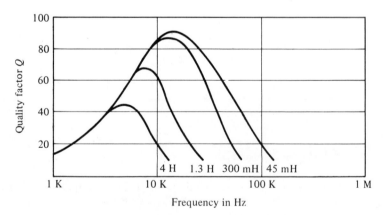

Q as a function of frequency: Microtran
QGL series toroidal inductors

Fig. 3–11 Q as a function of frequency for Microtran QGL series toroidal inductors. *(Courtesy of Microtran Co., Inc.)*

TABLE 3-2 Typical High-Q Iron-Core Inductors (TRW/UTC Transformers)

Series	Description	Size (in.)	Weight (oz.)	Inductance Range (0 dc)	mA dc (Maximum)	DCR (Ω/H)	Peak Q (Typical)
Subminiature							
ML	Laminated core, hipermalloy shielded case, straight pin terminals	$7/16$ x $1/2$ x $9/16$ h	0.20	0.15 to 60 H	12 to 0.2	85 to 150 Ω	22 @ 10 kHz
MM	Toroidal core, epoxy molded case, straight pin terminals	$7/16$ d x $1/4$ h	0.07	0.003 to 0.12 H	50 to 8	1300 Ω	60 @ 30 kHz
FE	Toroidal core, hermetically sealed molded case, flat construction	$15/16$ x $15/16$ x $1/2$ h	0.70	0.01 to 2 H	70 to 5	200 Ω	125 @ 8 kHz
Miniature							
FI	Toroidal core, hermetically sealed molded case, flat construction	$13/16$ x $13/16$ x $5/8$ h	1.50	0.04 to 4 H	50 to 5	100 Ω	150 @ 5 kHz
FO	Toroidal core, hermetically sealed molded case, flat construction	$17/8$ x $17/8$ x $7/8$ h	5.00	0.1 to 10 H	60 to 6	35 Ω	240 @ 3 kHz
MQM	Laminated core, hipermalloy shield, hermetically sealed case, two windings	$15/16$ d x $15/8$ h	5.00	2 to 600 H (series) 0.5 to 150 H (parallel)	—	10 Ω	40 @ 200 Hz

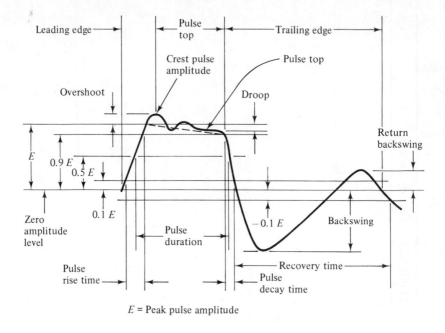

Fig. 3-12 Pulse waveform and parameters.

RISE TIME: The interval between 10% and 90% of the total amplitude. It is determined primarily by the high-frequency response of the transformer, but rise time may also be affected by the value of the load resistor—larger values decreasing the rise time.

PULSE WIDTH: The interval between the 50% amplitude points. The pulse width is determined primarily by the low-frequency response of the transformer and by the amount of droop that can be tolerated; the wider the pulse, the more droop.

DROOP: The displacement of the pulse amplitude; it is dependent mainly upon the low-frequency response of the transformer.

DECAY TIME: That portion of the wave from the 50% amplitude level of the trailing edge to the 0 or reference level.

BACKSWING: That portion of the wave that extends below the 0 level. The decay and backswing are dependent upon the open-circuit inductance, distributed capacity, and the load resistance. Decay is the action of the inductance discharging into the distributed capacity and the load resistance; backswing is the discharging of the distributed capacity into the open-circuit inductance and the load resistance. The resulting dampened ringing waveform may be reduced by lowering the value of the load resistor and thus dampening this exchange of energy.

Impedance Matching. Maximum transfer of power is realized when the input and output impedances are matched to their respective loads. It is not uncommon, however, to mismatch these impedances for some application. By reducing the secondary load resistance, the backswing would be decreased because of more dampening, the droop would decrease, the power transfer would decrease, and the rise time would be increased. A mismatch of as much as 10%, however, would be difficult to detect in any of these characteristics.

In a pulse-generating application, the characteristics of the other circuit components have a lot to do with the shape of the pulse and whether the circuit works at all.

Behavior of pulse transformers is not totally predictable. It is often necessary to buy a prototype and then find out by trial and error whether it can be made to work in a blocking oscillator circuit; if it doesn't work, it might be the transformer, your circuit, or both, and the best next step is to try a different transformer. Pulse transformers from two manufacturers, having identical catalog specifications, are not necessarily interchangeable in a blocking oscillator circuit.

Pulse transformers are also used to trigger lasers and thyristors, to isolate computer circuits, and to match impedances or voltages between pulse-forming circuits and microwave tubes.

3.6 WHO MAKES WHAT

Table 3–3 lists some of the major transformer manufacturers.

SUGGESTED READINGS

Hogan, C. Lester, "Ferrites," *Scientific American*. June 1960, pp. 92–104.

Langford-Smith, F., *Radiotron Designer's Handbook*. RCA, Harrison, N. J., 4th edition, 1952, pp. 140–58; 199–253; 429–80.

Low-Power Pulse Transformers, Engineering Bulletin No. 40,000A. Sprague Electric Company, North Adams, Mass., 1965.

Toroidal Inductor Application Guide. Microtran Company, Inc., Valley Stream, N. Y., April 1963.

TABLE 3-3 Transformer and Inductor Manufacturers

	James Millen	J. W. Miller	Microtran	Nytronics	Pico	Sprague	Stancor	Thordarson Meissner	TRW/UTC
Iron Core Transformers and Inductors (full line)							x	x	x
Miniature, Subminiature, and Ultraminiature Transformers and Inductors			x		x		x	x	x
Pulse Transformers					x	x		x	x
Radio Frequency Inductors	x	x		x					

CATALOGS

James Millen Manufacturing Co., Inc., Malden, Mass.

J. W. Miller Division of Bell Industries, Compton, Calif.

Pico Electronics, Inc., Mt. Vernon, N. Y.

UTC Transformers (TRW), New York, N. Y.

Thordarson Meissner, Mt. Carmel, Ill.

CHAPTER FOUR

Fixed Resistors

Resistors are circuit elements having the function of introducing electrical resistance into the circuit. There are three basic configurations: Fixed, rheostat, and potentiometer. A *fixed resistor* (Fig. 4–1) is a two-terminal resistor whose electrical resistance is constant (more or less). A *rheostat* is a resistor that can be changed in resistance value without opening the circuit to make the adjustment. A *potentiometer* is an adjustable resistor with three terminals, one at each end of the resistive element and the third movable along its length.

So much for basics. Fixed resistors are also made with adjustable taps so that their in-circuit resistance value can be changed, however the circuit is liable to be opened while doing so. Almost all rheostats are manufactured today with three terminals, making them in fact potentiometers, but they are still called rheostats and are used primarily as current-limiting devices. They can be used as potentiometers to provide at the tap a voltage that is a fraction of the voltage across the ends of the element, but in applications involving high power levels or long operating hours, semiconductor devices and variable autotransformers (ac only) are more economical. Potentiometers can be used as rheostats; the wide range of potentiometers made today are mostly for low-current applications.

Fixed and adjustable resistors are covered in this chapter. Potentiometers and rheostats are discussed in Chapter 5.

There are three basic types of fixed resistors: Carbon composition, in which the resistor element is a slug of resistive material; wirewound, in which the resistive element is a length of wire wound

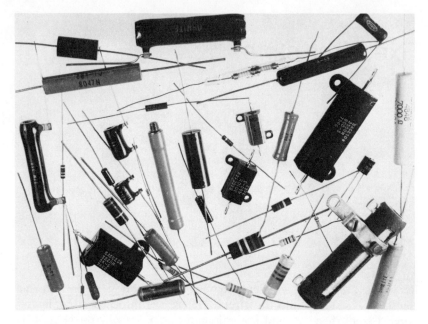

Fig. 4–1 Resistors.

in one or more layers on an insulating core; and film, in which the resistive element is a relatively thin layer of carbon, metal oxide, or metal, painted or deposited on a ceramic or glass tube or rod. (See Fig. 4–2.)

For convenience, resistors can also be classed as either small, where their electrical performance in the circuit is the primary concern, or power wirewound, where the primary function of the resistor is power dissipation. Resistors may also be classified by resistance tolerance:

General-Purpose	±20% to ±5%
Semi-precision	±5% to ±1%
Precision	±1% to ±0.1%
Ultra-precision	±0.1% to ±0.01%

The selection of a resistor for any application requires the consideration of numerous factors: Physical size, shape, method of mounting and connection into the circuit, resistance value, power dissipation, overload handling capability, reliability, and changes in nominal resistance with frequency, applied voltage, load-life, environmental conditions, and age. Although resistors are inexpensive

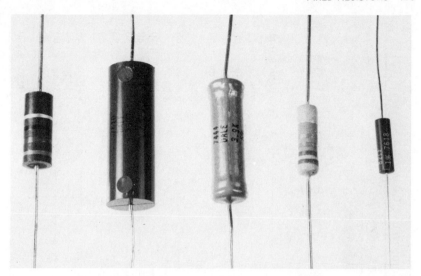

Fig. 4–2 One-watt resistors, shown full size. From left: Carbon composition, epoxy molded metal film, epoxy conformal coated carbon film, ceramic coated metal film, molded wirewound. *(Courtesy of Ohmite Mfg. Co., Dale Electronics Corp., Corning Glass Works.)*

compared to other circuit components, they are usually used in vast quantities, and initial cost must also be considered as an important factor.

Table 4–1 provides a general classification of fixed resistors.

Temperature Derating. In a circuit, a resistor converts electrical energy to heat. Under a constant electrical load, all resistors will attain a steady-state temperature at which the heat losses by radiation, convection, and conduction will equal the heat equivalent of the electrical input. The portion of the total heat loss by each mode varies with different resistor configurations and materials and with mounting.

Power ratings for resistors are established under a variety of industrial standards and MIL (Military) Specs. For high-power wirewound resistors, the various methods produce similar ratings. For small resistors, industrial and MIL ratings differ significantly.

Resistors do not heat evenly. There is a hot spot, usually somewhere near the center. All rating schemes are based on limiting this hot-spot temperature to some design value, and this maximum varies widely for different types of resistors. Operating at an ambient temperature the same as the hot-spot temperature, the resistor, regardless of its nominal power rating, can safely dissipate zero watts.

TABLE 4-1 Available Resistance Values and Comparative Cost

	Type	General-Purpose	Semi-precision	Precision	Ultra-precision	Power Ratings	Resistance Range (Standard)	Cost: Each in Small Quantity
	Carbon composition	+				1/8 to 2 w		$0.04 to $ 0.11
Film	Cermet		+			1/8 to 2 w		0.08 to 0.22
	Carbon film		+			1/2 to 2 w		1.00 to 2.00
	Metal oxide film			+	+	1/10 to 2 w		1.00 to 2.00
	Metal film			+	+	1/20 to 2 w		0.40 to 0.80
Small Axial-Lead Wirewound	Phenolic	+				1 to 2 w		0.16
	Ceramic shell	+				2 to 22 w		0.40 to 0.60
	Flameproof	+				1 to 10 w		1.00
	Vitreous enamel	+				1 to 11 w		0.35 to 1.10
	Silicone ceramic		+	+		1 to 15 w		1.10 to 4.40
	Silicone coated	+		+		0.25 to 15 w		0.60 to 1.50
	Bobbin			+		0.10 to 1 w		2.20 to 55.30
	Heat-sinked		+	+		7.50 to 100 w		2.00 to 6.00
Power Wirewound	Tubular	+	+			3 to 225 w		0.80 to 5.00
	Thin	+	+			10 to 55 w		1.50 to 4.00
	Tubular adjustable	+	+			12 to 225 w		1.50 to 7.00
	High-voltage/high-resistance	+	+			1 to 90 w		4.00 to 15.00

Resistance Range (Standard) scale markings: $.1\ \Omega$ | $1\ \Omega$ | $10\ \Omega$ | $100\ \Omega$ | $1\ k\Omega$ | $10\ k\Omega$ | $100\ k\Omega$ | $1\ M\Omega$ | $10\ M\Omega$ | $100\ M\Omega$ | $1000\ M\Omega$

At some specified lower maximum ambient temperature, in the range of 25°C to 125°C, a resistor can dissipate its full rated wattage and not exceed its hot-spot temperature. This temperature is variously called the "Maximum Power Rating," "Full Rating," or "Free Air Rating." The resistor is assumed to be operating in free air, but this benign condition is seldom met in practice, and derating to some fraction of the nominal rating is a normal design procedure.

The resistor can be operated above the maximum power rating temperature; but as this temperature is fixed, any increase in the ambient temperature above the rated power temperature subtracts from the available temperature rise, and the wattage must then be derated, generally on a straight-line basis reaching zero at the hot-spot temperature (Fig. 4–3).

Enclosures limit convection and radiation heat transfer. The walls of the enclosure also introduce a thermal barrier between the air contacting the resistor and outside cooling air. Size, shape, orientation, ventilating openings, wall thickness, material, and finish of enclosures all affect the temperature rise of the enclosed resistor.

Resistors mounted close to each other have higher hot-spot temperatures for a given wattage because of the heat received by radiation from each other.

The amount of heat that air can absorb varies with the density and therefore with the altitude. Above 100,000 feet, the air is so rare that the resistor heat transfer is by radiation and conduction only.

Forced air circulation removes more heat than natural convection and permits an increased watt rating. Liquid cooling and special heat-sink conduction mountings can increase the rating.

The only accurate method of determining the required derating is to measure the temperature rise with the resistor installed in the equipment.

It is also sometimes desirable to operate a resistor at a fraction of the nominal rating in order to keep the temperature rise low to protect adjacent heat sensitive components, to hold the resistance value stable with changing load, or to insure maximum life.

Marking. Fixed resistors are piece-marked in several ways: Color coding, straight numerical value, easily translated numerical codes, and one numerical code that is incomprehensible without a cheat-sheet.

Carbon composition, phenolic wirewound, and most semi-precision film resistors are marked with three or four painted colored bands (Fig. 4–4). Reading left to right, starting with the band nearest

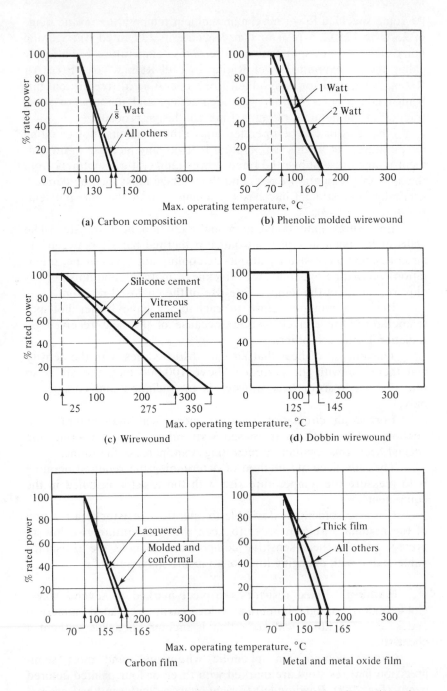

Fig. 4-3   Resistor temperature derating (industrial ratings shown).

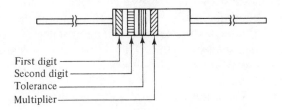

(a) Color code for general-purpose industrial resistors

Wide first band denotes
wirewound resistor

Wide first band and blue
fifth band denotes flame-
proof wirewound resistor

(b) Color code for general-purpose wirewound resistors

Color	Significant digit	Multiplier	Tolerance
Black	0	1	—
Brown	1	10	—
Red	2	100	—
Orange	3	1000	—
Yellow	4	10,000	—
Green	5	100,000	—
Blue	6	1,000,000	—
Violet	7	10,000,000	—
Gray	8	—	—
White	9	—	—
Gold	—	—	± 5%
Silver	—	—	±10%
No color	—	—	±20%

Example:

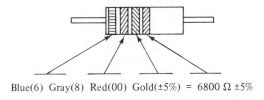

Blue(6) Gray(8) Red(00) Gold(±5%) = 6800 Ω ±5%

Fig. 4-4   EIA/MIL resistor color codes for ±5%, ±10% and ±20% resistors.

one end, the first two bands give two digits in the resistance value, the third band indicates the number of following zeros, and the fourth band indicates the tolerance. The absence of a fourth band indicates a tolerance of ±20%. This basic code, originally called the RMA (Radio Manufacturers Association) code is now called the EIA/MIL code (Electronic Industries Association/Military).

Variations in the width and additional bands further identify resistors. A wide first band indicates a wirewound resistor that could otherwise be indistinguishable from a carbon composition resistor. A wide first band and a fifth blue band indicate a wirewound flameproof resistor. Resistors having tolerances to 1% are identified by a fifth wide band spaced off at the right end of the resistor. MIL resistors may have a wide white band at the right end which denotes solderable leads (Fig. 4–5).

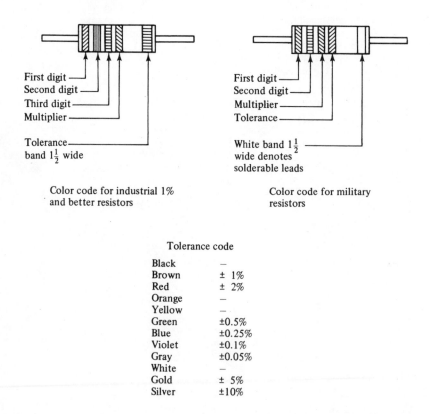

Color code for industrial 1% and better resistors

Color code for military resistors

Tolerance code

Black	—
Brown	± 1%
Red	± 2%
Orange	—
Yellow	—
Green	±0.5%
Blue	±0.25%
Violet	±0.1%
Gray	±0.05%
White	—
Gold	± 5%
Silver	±10%

Fig. 4–5 Color codes for industrial ±1% and better resistors and MIL-SPEC resistors.

Most industrial wirewound resistors are marked with their resistance value and tolerance; however, power-type wirewound resistors to MIL–R–26 are marked with a complex code that takes in resistance value, temperature coefficient, tolerance, and the temperature derating characteristics. To translate the coding, you need a copy of the MIL Spec.

Semi-precision, precision, and ultra-precision film and wirewound resistors (not power types) may be marked with either military or commercial alphanumeric codes. In all the military codes (Figs. 4–6 and 4–7), the first group of two or three letters identify the type of resistor (fixed film, fixed resistor, high stability, etc.) and the applicable MIL Spec. The first group of two numbers identify the body size and power rating in either of two numerical series (these numbers are also used in commercial marking). The next letter or letters can provide a variety of information — power rating temperature, temperature coefficient, solderable leads, etc., depending on the controlling MIL Spec. Resistance is denoted by the right-hand group of numbers; if three numbers, the first two give digits in the resistance value and the third gives the number of following zeros; if four numbers, there are three digits in the resistance value. The last letters in the code give tolerance and failure rate. Keeping in mind the layout of the MIL codes, it is simple to determine the resistance and tolerance of commercially marked resistors, but you will need the vendor's catalog to further identify the characteristics of the part.

Resistance Values. Except for some commercial power wirewound resistors, and special orders, all general-purpose resistors are made with resistance values in accordance with standard EIA/MIL values (Tables 4–2 and 4–3 on pages 114–16).

4.1 DEFINITIONS

CURRENT NOISE: The low-frequency noise caused by current flow through the nonhomogenous material of a carbon resistor. Noise power varies inversely with frequency and is generally not measurable above 100 kHz.

PURCHASE RESISTANCE: The resistance value marked or coded on a resistor.

PURCHASE TOLERANCE: The permissible resistance deviation from the value marked or coded on a resistor, applicable only at the time of purchase.

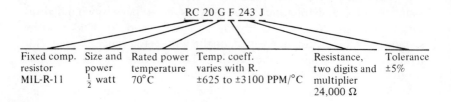

| Fixed comp. resistor MIL-R-11 | Size and power $\frac{1}{2}$ watt | Rated power temperature 70°C | Temp. coeff. varies with R. ±625 to ±3100 PPM/°C | Resistance, two digits and multiplier 24,000 Ω | Tolerance ±5% |

MIL-R-11 Resistor part numbers

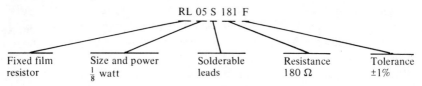

| Fixed film resistor | Size and power $\frac{1}{8}$ watt | Solderable leads | Resistance 180 Ω | Tolerance ±1% |

MIL-R-22684 Resistor part numbers (all have ± 200 PPM/°C temp. coeff.)

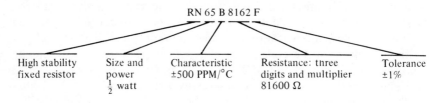

| High stability fixed resistor | Size and power $\frac{1}{2}$ watt | Characteristic ±500 PPM/°C | Resistance: three digits and multiplier 81600 Ω | Tolerance ±1% |

MIL-R-10509 Resistor part numbers

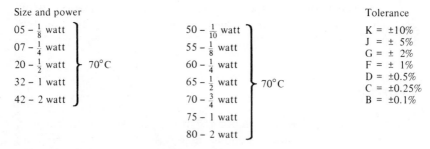

Size and power

| 05 – $\frac{1}{8}$ watt |
| 07 – $\frac{1}{4}$ watt |
| 20 – $\frac{1}{2}$ watt | 70°C |
| 32 – 1 watt |
| 42 – 2 watt |

| 50 – $\frac{1}{10}$ watt |
| 55 – $\frac{1}{8}$ watt |
| 60 – $\frac{1}{4}$ watt |
| 65 – $\frac{1}{2}$ watt | 70°C |
| 70 – $\frac{3}{4}$ watt |
| 75 – 1 watt |
| 80 – 2 watt |

Tolerance

K = ±10%
J = ± 5%
G = ± 2%
F = ± 1%
D = ±0.5%
C = ±0.25%
B = ±0.1%

Note: The MIL-R-26 (fixed wirewound resistors, power type) numbering system is complex and incomprehensible without a copy of the MIL spec.

Fig. 4–6   MIL-SPEC part numbering for carbon composition and film resistors.

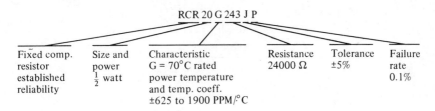

Fixed comp. | Size and | Characteristic | Resistance | Tolerance | Failure
resistor | power | G = 70°C rated | 24000 Ω | ±5% | rate
established | ½ watt | power temperature | | | 0.1%
reliability | | and temp. coeff. | | |
| | ±625 to 1900 PPM/°C | | |

MIL-R-39008 Resistor part numbers

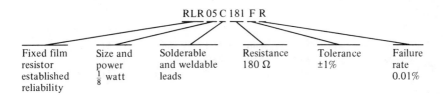

Fixed film | Size and | Solderable | Resistance | Tolerance | Failure
resistor | power | and weldable | 180 Ω | ±1% | rate
established | ⅛ watt | leads | | | 0.01%
reliability | | | | |

MIL-R-39017 Resistor part numbers

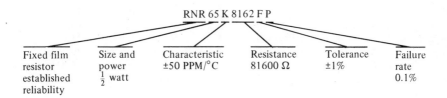

Fixed film | Size and | Characteristic | Resistance | Tolerance | Failure
resistor | power | ±50 PPM/°C | 81600 Ω | ±1% | rate
established | ½ watt | | | | 0.1%
reliability | | | | |

MIL-R-55182 Resistor part numbers

Characteristic (RN)

B = ±500 PPM/°C
C = ± 50 PPM/°C
D = ±200 or ±500 PPM/°C
E = ± 25 PPM/°C
F = ± 50 PPM/°C

Characteristic (RNR)

H = ± 50 PPM/°C
J = ± 25 PPM/°C
K = ±100 PPM/°C

Failure rate
per 1000 hours
(60% confidence)

M = 1.0%
P = 0.1%
R = 0.01%
S = 0.001%

Fig. 4-7 MIL-SPEC part numbering for established-reliability carbon composition, film, and wirewound resistors.

TABLE 4-2   Standard EIA/MIL Resistance Values for General-Purpose Resistors

A	±20% and ±10% Values						
1.0†	10	100	1000	10 k	100 k	1.0 M	10 M
1.2†	12	120	1200	12 k	120 k	1.2 M	12 M
1.5†	15	150	1500	15 k	150 k	1.5 M	15 M
1.8†	18	180	1800	18 k	180 k	1.8 M	18 M
2.2†	22	220	2200	22 k	220 k	2.2 M	22 M
2.7	27	270	2700	27 k	270 k	2.7 M	
3.3	33	330	3300	33 k	330 k	3.3 M	
3.9	39	390	3900	39 k	390 k	3.9 M	
4.7	47	470	4700	47 k	470 k	4.7 M	
5.6	56	560	5600	56 k	560 k	5.6 M	
6.8	68	680	6800	68 k	680 k	6.8 M	
8.2	82	820	8200	82 k	820 k	8.2 M	

B	±5% Values						
1.0†	10	100	1000	10 k	100 k	1.0 M	10 M
1.1†	11	110	1100	11 k	110 k	1.1 M	11 M
1.2†	12	120	1200	12 k	120 k	1.2 M	12 M
1.3†	13	130	1300	13 k	130 k	1.3 M	13 M
1.5†	15	150	1500	15 k	150 k	1.5 M	15 M
1.6†	16	160	1600	16 k	160 k	1.6 M	16 M
1.8†	18	180	1800	18 k	180 k	1.8 M	18 M
2.0†	20	200	2000	20 k	200 k	2.0 M	20 M
2.2†	22	220	2200	22 k	220 k	2.2 M	22 M
2.4†	24	240	2400	24 k	240 k	2.4 M	
2.7	27	270	2700	27 k	270 k	2.7 M	
3.0	30	300	3000	30 k	300 k	3.0 M	
3.3	33	330	3300	33 k	330 k	3.3 M	
3.6	36	360	3600	36 k	360 k	3.6 M	
3.9	39	390	3900	39 k	390 k	3.9 M	
4.3	43	430	4300	43 k	430 k	4.3 M	
4.7	47	470	4700	47 k	470 k	4.7 M	
5.1	51	510	5100	51 k	510 k	5.1 M	
5.6	56	560	5600	56 k	560 k	5.6 M	
6.2	62	620	6200	62 k	620 k	6.2 M	
6.8	68	680	6800	68 k	680 k	6.8 M	
7.5	75	750	7500	75 k	750 k	7.5 M	
8.2	82	820	8200	82 k	820 k	8.2 M	
9.1	91	910	9100	91 k	910 k	9.1 M	

† Industrial grade only.

TABLE 4-3   Standard EIA/MIL Resistance Values for Ultra-Precision, Precision and Semi-Precision Resistors†

*	±1%	±2%	*	±1%	±2%	*	±1%	±2%
1.00	1.00	1.0	1.58	1.58		2.49	2.49	
1.01			1.60			2.52		
1.02	1.02		1.62	1.62	1.6	2.55	2.55	
1.04			1.64			2.58		
1.05	1.05		1.65	1.65		2.61	2.61	
1.06			1.67			2.64		
1.07	1.07		1.69	1.69		2.67	2.67	
1.09			1.72			2.71		2.7
1.10	1.10	1.1	1.74	1.74		2.74	2.74	
1.11			1.76			2.77		
1.13	1.13		1.78	1.78	1.8	2.80	2.80	
1.14			1.80			2.84		
1.15	1.15		1.82	1.82		2.87	2.87	
1.17			1.84			2.91		
1.18	1.18		1.87			2.94	2.94	
1.20			1.89			2.98		
1.21	1.21	1.2	1.91	1.91		3.01	3.01	3.0
1.23			1.93			3.05		
1.24	1.24		1.96	1.96		3.09	3.09	
1.26			1.98			3.12		
1.27	1.27		2.00	2.00	2.0	3.16	3.16	
1.29			2.03			3.20		
1.30	1.30	1.3	2.05	2.05		3.24	3.24	
1.32			2.08			3.28		
1.33	1.33		2.10	2.10		3.32	3.32	3.3
1.35			2.13			3.36		
1.37	1.37		2.15	2.15		3.40	3.40	
1.38			2.18			3.44		
1.40	1.40		2.21	2.21	2.2	3.48	3.48	
1.42			2.23			3.52		
1.43	1.43		2.26	2.26		3.57	3.57	
1.45			2.29			3.61		
1.47	1.47		2.32	2.32		3.65	3.65	3.6
1.49			2.34			3.70		
1.50	1.50	1.5	2.37	2.37	2.4	3.74	3.74	
1.52			2.40			3.79		
1.54	1.54		2.43	2.43		3.83	3.83	
1.56			2.46			3.88		

*±0.1%, ±0.25%, and ±0.5%.

† Values are given for one decade. All other decades are the same with appropriate multiplier. For each type of resistor, values are subject to minimum and maximum limits.

TABLE 4-3   *Continued*

*	±1%	±2%	*	±1%	±2%	*	±1%	±2%
3.92	3.92	3.9	5.36	5.36		7.32	7.32	
3.97			5.42			7.41		
4.02	4.02		5.49	5.49		7.50	7.50	7.5
4.07			5.56			7.59		
4.12	4.12		5.62	5.62	5.6	7.68	7.68	
4.17			5.69			7.77		
4.22	4.22		5.76	5.76		7.87	7.87	
4.27			5.83			7.96		
4.32	4.32	4.3	5.90	5.90		8.06	8.06	
4.37			5.97			8.16		
4.42	4.42		6.04	6.04		8.25	8.25	8.2
4.48			6.12			8.35		
4.53	4.53		6.19	6.19	6.2	8.45	8.45	
4.59			6.26			8.56		
4.64	4.64	4.7	6.34	6.34		8.66	8.66	
4.70			6.42			8.76		
4.75	4.75		6.49	6.49		8.87	8.87	
4.81			6.57			8.98		
4.87	4.87		6.65	6.65		9.09	9.09	9.1
4.93			6.73			9.20		
4.99	4.99		6.81	6.81	6.8	9.31	9.31	
5.05			6.90			9.42		
5.11	5.11	5.1	6.98	6.98		9.53	9.53	
5.17			7.06			9.65		
5.23	5.23		7.15	7.15		9.76	9.76	
5.30			7.23			9.88		

*±0.1%, ±0.25%, and ±0.5%.

RESISTIVITY: The measure of the resistance of a material to current flow through it, expressed, in resistor technology, in ohms per square.

TEMPERATURE COEFFICIENT OF RESISTANCE: The ratio of the change in resistance to the purchase resistance for a change in temperature, usually expressed in parts per million per degree Celsius (ppm/°C).

THERMAL NOISE: The noise caused by the thermal agitation of the current carriers in a resistor. It is dependent on the absolute temperature of the resistor, but independent of the material. Also called *Johnson noise*.

VOLTAGE COEFFICIENT OF RESISTANCE (VC): The percentage change in resistance per volt (dc) from the purchase resistance of a resistor, usually given as percent per degree Celsius.

4.2 CARBON COMPOSITION RESISTORS

Molded, fixed composition resistors are intended for general-purpose use in electronic equipment and are unequaled for uniformity, predictable performance, and freedom from catastrophic failure. They are able to withstand higher voltages than film resistors of standard configuration, have very low inductance and capacitance, can tolerate rough handling during installation, and are inexpensive.

Construction. The resistive element in a carbon composition resistor is a solid rod with leads on the ends (Fig. 4–8). The material is a mixture of finely ground carbon or graphite, an inert nonconducting filler, usually silica, and a synthetic resin binder. The mix of carbon and filler is adjusted to produce different resistances. The more carbon, the lower the resistance. A charge of this mixture, plus leads and insulating material, is compressed and bonded under pressure, and is then cured in a controlled bake cycle. In an alternate process, pressure and heat are combined in a single step. The resistor is then vacuum-impregnated with a sealant to provide a barrier to moisture penetration.

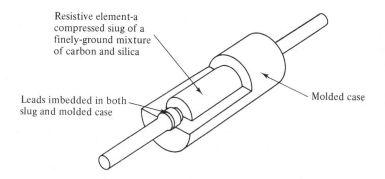

Resistive element-a
compressed siug of a
finely-ground mixture
of carbon and silica

Leads imbedded in both
slug and molded case

Molded case

Fig. 4–8   Carbon composition resistor construction.

Performance Characteristics.   (See Table 4–4.) Resistance change over load–life is permanent and usually negative. Load–life stability can be improved with derating or by forced cooling to keep the body temperature below 100°C. Long-time storage at room ambient causes negligible resistance change.

Noise in a carbon composition resistor is a combination of thermal and current noise. The current noise is due to the particle-to-particle conduction. (In old-time radio, this effect was put to use in

TABLE 4-4 Carbon Composition Resistors

	All Sizes				
Resistance Range	2.7 Ω to 22 MΩ				
Purchase Tolerance	±20%, ±10%, ±5%				
Operating Temperature Range	−55°C to +150°C				
Temperature Coefficient	±1200 ppm/°C				
Voltage Coefficient	−0.5 v/vdc				
Noise	$\left(2 + \log \dfrac{R}{1000}\right)$ v/vdc				
Capacitance, Typical	0.25 pF				
Inductance, Typical	Nil				
Power Rating	1/8 w	1/4 w	1/2 w	1 w	2 w
Rated Power Temperature	130°C	150°C	150°C	150°C	150°C
Working Voltage, dc	150 v	250 v	350 v	500 v	750 v

carbon microphones by voice vibration of loosely spaced particles.) For low-noise applications, use a wirewound or a film resistor.

Humidity has more effect on high-value resistors than on low values. Increased moisture content causes the resistance to go up; the change is temporary and can be corrected by baking out.

The inductance of carbon composition resistors can usually be ignored. Shunt capacity is a function of resistance value and resistor configuration and is usually less than 0.5 pF.

Dc resistance and ac resistance are equal up to 20 kHz; then with increasing frequency, ac resistance drops off, largely due to distributed capacitance (it drops off faster for high resistance values than for low). For any specific resistor slug configuration, the ratio of ac resistance to the dc resistance is approximately constant for a fixed value of the product of operating frequency and dc resistance value.

Put another way, for a typical resistor element configuration, resistance at 1 MHz could be for different values:

10 MΩ resistor	3.8 MΩ
1 MΩ "	780 kΩ
100 kΩ "	99 kΩ
10 kΩ "	10 kΩ

The best high-frequency characteristics in a carbon composition resistor are attained when the cross-sectional area-to-length ratio of the element is small.

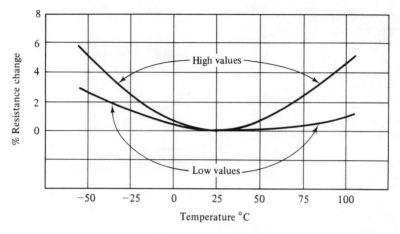

Fig. 4–9   Carbon composition resistor resistance change with temperature.

Resistance increases at high and low temperatures, more so for high values than for low values. See Fig. 4–9.

4.3 WIREWOUND RESISTORS

Wirewound resistors (Fig. 4–10) are manufactured in a greater variety of resistance values, tolerances, power ratings, and physical configurations than the other types of resistors. They replace carbon composition resistors in low-resistance or low-noise applications or when higher power-handling capability is required in the same physical space. Precision wirewound resistors are chosen over film resistors when ability to handle transient or repetitive pulses is needed. Small, axial, lead wirewound resistors are ideal for high density circuits and PC board mounting. Molded varieties may be clip-mounted to withstand vibration and shock or to improve heat dissipation. For any resistor application requiring a nominal power rating above 2 watts, you don't have much other choice.

Construction.   Most wirewound resistors are wound with a single-layer spiral element of high resistance wire (Fig. 4–11). Noninductive varieties and multilayer bobbin wound resistors are also made. The fragile windings are protected against mechanical and environmental hazards by dipped, sprayed, molded, or rigid covers of high-temperature silicone, inorganic cement, vitreous enamel, plastic, or ceramic materials.

OHMITE TYPE 200
"BROWN DEVIL" (12 W)
VITREOUS ENAMEL

OHMITE TYPE 210
"DIVIDOHM" (12 W)
VITREOUS ENAMEL

OHMITE SERIES 22
(11 W) VITREOUS ENAMEL

OHMITE SERIES 884
RITEOHM (11 W) MOLDED
SILICONE CERAMIC

OHMITE SERIES 99
(W) MOLDED VITREOUS
ENAMEL

SPRAGUE KOOLOHM
(W) CERAMIC TUBE
ENCASED

IRC TYPE PW (10 W)
CERAMIC FIREPROOF

SAGE TYPE C (11 W)
METAL SHEATHED

SAGE TYPE M (8 W)
HEAT SINKED

OHMITE TYPE 57
"CERON" (5 W)
FLAME PROOF

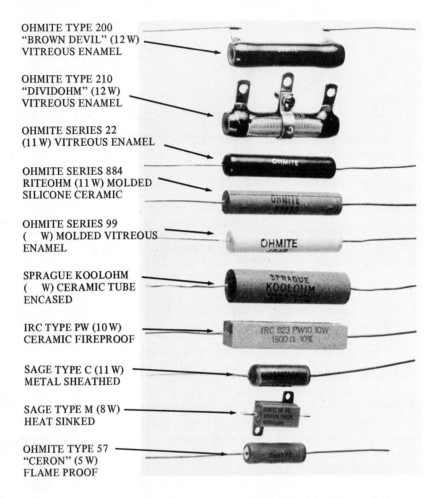

Fig. 4–10 Wirewound resistors, shown half size. *(Courtesy of Nytronics, Inc., Ohmite Mfg. Co., TRW/IRC Potentiometers.)*

These resistors vary in size according to power ratings. There also can be differences in size for the same power rating between classes of wirewound resistors (Fig. 4–12); differences increase with power rating. Maximum hot-spot temperatures vary widely, and not all wirewound resistors are nominally rated at 25°C.

Phenolic Molded. These low-cost one-watt and two-watt wirewound resistors have the same body sizes as half-watt and one-watt carbon composition resistors. They are useful in general-purpose applications requiring lower-range resistance values than are

Resistance wirewound
on ceramic rod substrate

Resistance wire
welded to end caps

Leads welded
to end caps

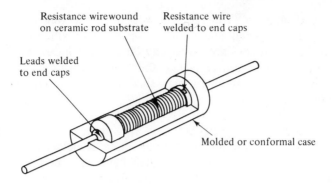

Molded or conformal case

Fig. 4–11   Wirewound resistor construction.

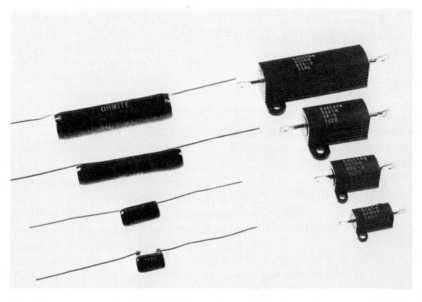

Fig. 4–12   Power wirewound resistors, shown half-size. Left: Vitreous enamel coated 3 watt, 5 1/4 watt, 12 watt and 20 watt "Brown Devil" wirewound resistors *(Courtesy of Ohmite Mfg. Co.)*. Right: 8 watt, 14 watt, 25 watt and 50 watt metal-sheathed wirewound resistors *(Courtesy of Nytronics, Inc.)*.

available in carbon composition resistors, double wattage in the same package size, or low current noise. Resistance elements are wound on fiberglass cores and encased in molded phenolic shells.

Ceramic Shell.   The ceramic covering of these low-cost in-dustrial resistors provides good physical protection and excellent insulation. The single-layer element is wound on a ceramic core with leads welded. The shell openings are sealed with silicone cement. Although these resistors are physically larger than industrial tubular resistors of the same power rating, they have advantages of being more easily PC-board mounted, heat-sinked with mounting clips, and inserted by machine.

Flameproof. These resistors are constructed of inorganic materials and will not ignite under any magnitude of electrical overload, nor will they ignite on exposure to open flame. Under excess load, the winding opens cleanly. Applications for these resistors include home entertainment, medical, communications, and data processing equipment.

Vitreous Enamel. These wirewound resistors have high heat dissipation capabilities. The glassy surface provides an impervious shield against moisture and resists solvents. The vitreous enamel coating may be either molded or conformal; if molded, the resistor may be mounted in a heat-sink clip.

Silicone Ceramic. This material provides a tougher cover for wirewound resistors than vitreous enamel and is better able to withstand vibration and shock. The covering can be molded or conformal.

Silicone Coated. A miniature silicone-coated resistor that is rated 3 watts at 125°C has the same body size as a 1-watt carbon composition resistor, which at 125°C would have to be derated to 1/3 watt.

Table 4–5 provides a comparison of low-power general-purpose and semiprecision wirewound resistors.

Precision and Ultra-precision Wirewound. These resistors (Table 4–6) are made in molded, conformal-coated, and heat-sinked versions (see below). They are available in lower resistance values than other types of precision resistors (0.005Ω versus 10Ω), have lower noise, and can operate at higher temperatures, but their high-frequency use is limited by their inherently inductive construction.

Bobbin. A short time ago, a precision wirewound resistor meant a bobbin resistor—one with a multilayer winding of relatively low-resistance wire. Bobbin resistors are bulky, have low wattage ratings in the order of one watt maximum, and are expensive. How-

TABLE 4-5   Typical Low-Power General-Purpose and Semi-precision Wirewound Resistors

	General-purpose			Semi-precision	
Style	Phenolic molded	Vitreous enamel; molded, or conformal	Ceramic flameproof molded	Silicone-ceramic molded, conformal	Silicone-coated conformal
Example	TRW/IRC BW–20, BWH	Ohmite Series 22, 99	Ohmite Series 57 Ceron®	Ohmite Series 44, 88	TRW/IRC AS
Power Rating	1, 2 w	1 to 11 w	1 to 10 w	1 to 15 w	¼ to 15 w
Resistance Range, Low	0.24 Ω	0.1 Ω	1 Ω	0.1 Ω	0.1 Ω
Typical High	2400 Ω	200 kΩ	60 kΩ	300 kΩ	175 kΩ
Purchase Tolerance, Standard	±10%, ±5%	±5%	±5%	±5% to ±1%	±5% to ±0.05%
Operating Temperature, Maximum	+160°C	+350°C	+350°C	+275°C	+350°C
Rated Power Temperature, Maximum	50°C, 70°C	25°C	25°C	25°C	25°C
Temperature Coefficient	±200 ppm/°C	±30 ppm/°C	±20 ppm/°C	±20 ppm/°C	±20 ppm/°C

TABLE 4-6 Precision and Ultra-precision Wirewound Resistors†

	Molded	Silicone Coated	Aluminum Heat-sinked	Bobbin, Molded Epoxy
Power Rating	¼ to 10 w	¼ to 10 w	7½ to 100 w	1/10 to 1 w
Resistance Range Low	0.5 Ω	0.1 Ω	0.1 Ω	0.1 Ω
Typical High	250 kΩ	250 kΩ	100 kΩ	18 MΩ
Purchase Tolerance, Standard	±5%, ±3%, ±1%, ±0.5%, ±0.1%, ±0.05%	±5%, ±3%, ±1%, ±0.5%, ±0.1%, ±0.05%	±3%, ±1%, ±0.5%, ±0.25%, ±0.1%, ±0.05%	±1%, ±0.5%, ±0.25%, ±0.1%, ±0.05%
Operating Temperature, Maximum	275°C	275°C	275°C	145°C
Rated Power Temperature, Maximum	25°C	25°C	25°C	125°C
Temperature Coefficient	±20 ppm/°C	±50 ppm/°C	±50 ppm/°C, ±30 ppm/°C, ±20 ppm/°C	±20 ppm/°C to ±1 ppm/°C

† Also available in noninductive versions

ever, available resistances are higher than for other wirewound resistors, pairs can be matched to a tolerance of 0.001%, and temperature coefficients can be matched to ± 2 ppm/°C over limited temperature ranges.

Heat-Sinked. Wirewound resistors encased in finned aluminum housings can dissipate more power in the same volume than other types—even more if mounted on an aluminum chassis or plate.

Power Wirewound Resistors. These resistors are used mainly to dissipate relatively large amounts of electrical power in a small space. Precise resistance value is usually not a requirement. The most common construction consists of a hollow cylindrical ceramic tube core having band terminals around each end with a single-layer winding of resistance wire spiraled between. A conformal coating of either vitreous enamel or silicone ceramic cement is applied over the winding and bands for physical protection and electrical insulation. (See Table 4–7.)

Terminal lugs or tabs are brought out radially; some smaller-sized resistors have lead wires welded to the lugs, and the resistors are mounted by their leads in applications not subject to any vibration or shock. Brackets or through-bolts are used to mount the larger sizes.

Where space is limited, flat core (also called thin, strip, and oval core) can be substituted for the round core type. These resistors are constructed similarly to the tubular types except for shape—they are mounted by brackets and can be stacked; however, when stacked, they must be derated to limit the hot spot of the hottest one to the maximum allowed.

Both round and flat resistors are available in adjustable resistance versions. These have the side of the winding bared the length of the winding with adjustable connection made by means of a screw-tightened lug. These resistors are used when occasional adjustment to the resistance is required, to obtain nonstandard or tapped resistances, or to make voltage dividers. However, the wattage rating of the tapped resistor applies only when the entire resistor is in the circuit; at intermediate points, the wattage is reduced proportionally.

Noninductive Wirewound Resistors. A single-layer spiral-wound resistor by its configuration is an inductor with shunt capacitance. For most applications at frequencies up to 25 kHz, this is not a problem; at higher frequencies, noninductive resistors are usually required.

TABLE 4-7 High-Power Wirewound Resistors

	Ceramic Shell	Aluminum Heat-sinked	Tubular, Conformal Vitreous Enamel	Thin (Flat) Tubular, Conformal Vitreous Enamel	Adjustable Tubular, Conformal Vitreous Enamel	Tubular, Conformal Silicone Ceramic
Power Rating	2 to 22 w	7½ to 100 w	3 to 225 w	10 to 55 w	12 to 225 w	5 to 225 w
Resistance Range, Low	0.1 Ω	0.1 Ω	0.5 Ω	1 Ω	1 Ω	0.1 Ω
Typical High	70 kΩ	100 kΩ	250 kΩ	50 kΩ	100 kΩ	1.3 MΩ
Purchase Tolerance, Standard	±5%, ±10%	±5% to ±0.05%	±5%	±5%	±5%	±5%
Operating Temperature, Maximum	275°C, 350°C	275°C	350°C	350°C	350°C	275°C
Rated Power Temperature, Maximum	25°C	25°C	25°C	25°C	25°C	25°C
Temperature Coefficient	±260 ppm/°C	±20 ppm/°C	±260 ppm/°C	±260 ppm/°C	±260 ppm/°C	±20 ppm/°C

The noninductive equivalent for a single-layer resistor is usually a two-layer Ayrton-Perry winding. In this type of winding, two windings with the same number of turns are wound on the core in opposite directions and connected in parallel (Fig. 4–13). Not only do the inductances of the two windings cancel, but distributed capacitance is reduced. A noninductive resistor has about 1% of the inductance of a standard wound resistor of the same size. The effect on wirewound resistor characteristics is a function not only of inductance and capacitance, but also the resistance of the resistor. Below 750 Ω, the residual reactance is inductive; above 750 Ω, it is capacitive.

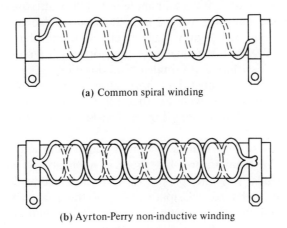

(a) Common spiral winding

(b) Ayrton-Perry non-inductive winding

Fig. 4–13 Ayrton-Perry non-inductive winding. Each of the two windings on the core must have twice the desired resistance. It is not necessary to use insulated wire.

Special Wirewound Types. Power wirewound resistors are made in a variety of additional styles to meet special requirements, but they must be purchased on special order. Although they are available only in a limited range of power ratings, you can, within limits, get exactly the resistance you want.

Ferrule types, used in transmitters and large power supplies, can be mounted in clips similar to those used for cartridge fuses. Edison screw-base resistors mount in light-bulb sockets. Noninductive dummy antenna resistors in standard matching resistances (52 Ω, 73 Ω, 300 Ω, and 600 Ω) with appropriate housings and connectors are made for tuning transmitters, etc., up to 30 MHz and 250 watts.

For low-resistance applications requiring high intermittent power ratings, resistors having a corrugated edge-wound ribbon are made with the element partially imbedded in vitreous enamel, or bare on the ceramic core, or with a bare round-wire element.

Tubular resistors can be ordered with one or more fixed taps (once a quite common configuration) or with both connection tabs at one end.

Pulse Rating.   Wirewound resistors have a pulse-handling capability far higher than their steady-state power rating would indicate. For example, a 10-Ω 5-watt resistor with a maximum operating temperature of 350°C can handle a 24,600-watt square pulse of a 1-millisecond duration, or a 24,600,000-watt pulse of a 1-microsecond duration. Although voltage overload ratings have not been established, 20,000 volts per inch of resistor length would be typical for a pulse of less than a 100-millisecond duration.

Calculations for repetitive short duration pulses must consider the average power as well as the pulse power. Noninductively wound resistors can handle higher power pulses than standard wirewound resistors.

In pulses of less than a 100-millisecond duration, the energy raises the wire temperature, but there is no heat loss in the core, coating, or leads. For long pulses, of up to a 5-second duration, the core, coating, and leads dissipate much of the heat, and the published 5 or 10 times the 5-second overload ratings can be used, corrected for any shorter duration. For example, a 1-second pulse overload rating would be 5 times the 5-second rating. For pulses longer than 100 milliseconds, voltage is limited to 2 to 3 times the maximum steady-state working voltage.

4.4 FILM RESISTORS

In all film resistors (Fig. 4–14), the element is a thin layer of resistive material applied to the surface of a ceramic or glass rod or tube. The resistive material can be a carbon dispersion, deposited carbon, tin oxide, or any of several metals. Depending on the material, the film may be formed by spraying, dipping, evaporation, sputtering, or pyrolytic cracking of gas. Resistance is dependent on the choice of material and the thickness of the coating.

Construction.   Before encapsulation, all except low-value film resistors are adjusted to final resistance by spiraling, which means cutting a helical groove around the cylinder, which simultaneously

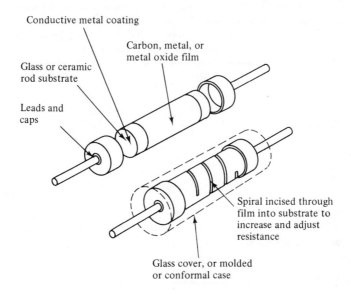

Conductive metal coating

Carbon, metal, or
metal oxide film

Glass or ceramic
rod substrate

Leads and
caps

Spiral incised through
film into substrate to
increase and adjust
resistance

Glass cover, or molded
or conformal case

Fig. 4–14   Film resistor construction.

increases the length of the resistive path between end terminals and reduces the width of the path. Three methods are used: Abrasive powder carried in a carefully aimed high velocity air stream, a thin high-speed grinding wheel, or a laser beam. The spiral groove is usually about 0.01 inch wide. The resistance of an uncut resistor can be increased up to 3000 times by spiraling. In all methods, the operation is automatic; the resistor is rotated while the cutter traverses the length of the resistor. Resistance is continuously monitored; when the desired value is reached, grooving is abruptly halted. Accuracies of ±0.01% are obtained.

Spiraling affects high-frequency performance, but not usually below 1 MHz. Above that frequency, capacitive reactance becomes a problem. Unspiraled low-value film resistors are usable in the 100 MHz region.

After spiraling, leads are attached and the resistor encapsulated.

Carbon Film Resistors.   Pure carbon can be deposited onto a ceramic rod by the high-temperature thermal decomposition of gaseous hydrocarbons. Careful control of the thickness can produce a wide range of resistances. It is possible to manufacture deposited carbon resistors up to around 1000 Ω without spiraling. (See Table 4–8.)

TABLE 4-8   Semi-precision and Precision Deposited
Carbon Film Resistors

		Semi-precision	Precision	
			Precision	
Style		Lacquer	Molded	Conformal
Power Rating		$\frac{1}{8}$, $\frac{1}{4}$, $\frac{1}{2}$ w	$\frac{1}{10}$, $\frac{1}{8}$, $\frac{1}{4}$, $\frac{1}{2}$, 1, 2 w	$\frac{1}{8}$, $\frac{1}{4}$, $\frac{1}{2}$, 1, 2 w
Resistance Range	Low	1 Ω	1 Ω	1 Ω
Typical	High	1 MΩ	100 MΩ	25 MΩ
Purchase Tolerance, Standard		±5%	±2%, ±1%, ±0.5%	±2%, ±1%, ±0.5%
Operating Temperature, Maximum		+155°C	+165°C	+165°C
Rated Power Temperature		+70°C	+70°C	+70°C
Temperature Coefficient		−200 ppm/°C to −500 ppm/°C	−200 ppm/°C to −500 ppm/°C	−200 ppm/°C to −500 ppm/°C

The big advantage of a carbon film resistor over a carbon composition resistor is better stability. Purchase tolerances range from ±5% to ±0.5%. Load–life resistance change is improved, and the voltage coefficient is negligible. The temperature coefficient depends on the thickness of the film and is negative. For the same power ratings, the body dimensions run larger than carbon composition resistors.

The thin fragile film has no ability to withstand electrical overloads of any magnitude. However, unlike the carbon composition resistor that bakes for a while before smoking, the carbon film resistor opens fast and clean, making it useful as a fuse.

Metal Film Resistors. The development of precision-deposited metal film resistors had the objectives of incorporating the advantages of both deposited carbon resistors and precision wirewound resistors without incurring the inherent disadvantages of either. Metal film resistors are used in applications that require a combination of long life under load and the best possible reliability and stability. Their ability to handle short-time overloads is superior to that of carbon film resistors but not as good as that of carbon composition or cermet resistors. Purchase tolerances range from ±5% to ±0.01%. (See Table 4–9.)

These resistors are made by coating a glass or ceramic substrate with a very thin film of a metal, metal alloy, or metal oxide. The materials and processes used vary from manufacturer to manu-

TABLE 4-9 Semi-precision, Precision, and Ultra-precision Metal Film Resistors

		Semi-precision		Precision			Ultra-precision
Resistive Element		Tin oxide	Cermet (thick film)	Metal film	Metal film	Tin oxide film	Metal film
Case		Glass	Molded phenolic	Molded epoxy	Conformal	Glass	Molded epoxy
Power Rating		$1/8$, $1/4$, $1/2$, 1, 2 w	$1/8$, $1/4$, $1/2$, 1, 2 w	$1/8$, $1/4$, $1/2$, $3/4$, 1, 2 w	$1/8$, $1/4$, $1/2$, 1 w	$1/10$, $1/8$, $1/4$, $1/2$, $3/4$, 1 w	$3/10$ w
Resistance Range Low		10 Ω	4.3 Ω	10 Ω	10 Ω	50 Ω	20 Ω
Typical High		2 MΩ	1.5 MΩ	158 MΩ	24 MΩ	1 MΩ	100 kΩ
Purchase Tolerance, Standard		±5%, ±2%, ±1%	±5%, ±2%	±1%, ±0.5%, ±0.25%, ±0.1%	±1%, ±0.5%, ±0.25%, ±0.1%	±1%, ±0.5%, ±0.25%, ±0.1%	±1%, ±0.5%, ±0.25%, ±0.1%, ±0.05%, ±0.02%, ±0.01%
Operating Temperature, Maximum		150°C	150°C	165°C	165°C	165°C	85°C
Rated Power Temperature, Maximum		70°C	70°C	70°C	70°C	70°C	—
Temperature Coefficient, Standard		±200 ppm/°C, ±100 ppm/°C	±250 ppm/°C, ±200 ppm/°C	±100 ppm/°C, ±50 ppm/°C, ±25 ppm/°C	±100 ppm/°C, ±50 ppm/°C, ±25 ppm/°C	±50 ppm/°C, ±25 ppm/°C	±5 ppm/°C, ±2 ppm/°C

131

facturer, and exact details are proprietary. The following processes are typical.

Nickel-Chromium Film.   In this, the industry's original metal film process, a metal film is vapor-deposited in a vacuum onto a ceramic rod substrate. The film contains nickel, chromium, and traces of other metals. The film thickness and proportion of nickel and chromium are controlled to obtain the desired resistance, which can be varied. A protective coating (silicon monoxide) is evaporated over the film. Thicknesses of as little as one-millionth of an inch are possible, and very low temperature coefficients can be obtained. After spiraling, the resistor is encapsulated in glass or molded or conformal plastic.

Tin-Oxide Film.   These resistors are made by vapor-depositing a film onto a solid glass rod substrate. The film is a semiconducting tin and tin oxide film doped with antimony to provide varying values of film resistivity. Film deposition is generally done by a high-temperature reaction of stannic chloride vapor and water vapor: An antimony-doped solution of stannic chloride is sprayed on a heated substrate; as the solution strikes the substrate, it is oxidized, and a conducting film is formed. The ends of the resistor rod are coated with a conducting material which is then fired to provide cohesion. The resistor blanks are helix-cut to obtain the desired final resistance, leads are attached, and the resistor is sealed in glass or coated with a moisture-resistant varnish.

Cermet.   These molded, thick-film, semi-precision resistors provide a combination of ruggedness and stability at close to the price of carbon composition resistors. The resistors are made by dipping a heat-conducting alumina rod substrate into a mixture of glass and metal alloys; the glaze is fused to the ceramic core at 1100°C. The resistance element is up to 100 times thicker than an oxide or evaporated metal film. This process produces a stable resistance element that is extremely hard and durable and relatively impervious to environmental extremes. Cermet resistors withstand overloads, but not as well as comparable wirewounds.

Film Resistor Design Tolerances.   Circuits must be designed to operate satisfactorily, taking into account all possible combinations of variations in component parameters allowed by component tolerances. (See Table 4–10.)

It is a bit of an error to assume that a resistor sold with a given

TABLE 4-10 Comparison of Small Precision and Ultra-Precision Resistors (Purchase Tolerances ±1% to ±0.01%)

	Wirewound	Wirewound (Bobbin)	Carbon Film	Metal, Metal Oxide Film
Resistance Range Low	0.005 Ω	0.1 Ω	10 Ω	10 Ω
High	500 kΩ	18 MΩ	125 MΩ	148 MΩ
Temperature Coefficients, Typical	±20 ppm/°C	±20 ppm/°C to ±1 ppm/°C	±500 ppm/°C	±10 ppm/°C
Standard Power Rating	$1/8, 1/4, 1/2,$ 1, 2 to 15 w	0.1 to 1.0 w	$1/10, 1/8, 1/4,$ $1/2,$ 1, 2 w	$1/8, 1/8, 1/4, 1/2,$ $3/4,$ 1, 2 w
Power Rating Temperature	25°C	125°C	70°C	70°C, 125°C
Operating Temperature Ranges	−55 to +275°C −55 to +350°C	−55 to +145°C	−55 to +165°C	−55 to +175°C
Stability	ΔR	ΔR	ΔR	ΔR
Temperature Cycling	±(0.2% + 0.05 Ω)	±0.07%	±0.3%	±0.1%
Low Temperature Operation	—	±0.02%	±0.05%	±0.05%
Short Time Overload	±(0.2% + 0.05 Ω)	±0.01%	±0.05%	±0.02%
Dielectric	±(0.1% + 0.5 Ω)	±0.00%	±0.01%	±0.01%
Effects of Soldering	—	—	±0.01%	±0.02%
Moisture Resistance	±(0.2% + 0.05 Ω)	±0.02%	±0.3%	±0.05%
Shock	±(0.1% + 0.05 Ω)	±0.00%	±0.1%	±0.01%
Vibration	±(0.1% + 0.05 Ω)	±0.01%	±0.1%	±0.01%
Load Life	±(0.5% + 0.05 Ω)	±0.09%	±0.1%	±0.05%

tolerance will remain within that tolerance forever. A 1000-Ω, ±1% resistor cannot be assumed to stay between 990 Ω and 1010 Ω over its operational life. That tolerance is only the *purchase tolerance*. A further allowance must be made for normal changes in the value of the part in use. The limits beyond which the part will not vary during its useful life make up its *design tolerance*. All of the following can affect the resistance of a resistor: Moisture, temperature cycling, short-time overloads, low- and high-temperature operation, soldering, handling, shock and vibration, hours of operation, and weeks spent on the shelf waiting for use. In other words, how close the resistor stays to the nominal value depends on the type of resistor, environment, application—and luck.

All resistors have temperature coefficients that must also be taken into consideration, and carbon resistors have a voltage coefficient. However, arithmetically adding together the purchase tolerance and all the maximum guaranteed deviations to arrive at a design tolerance is not necessary when using tight-tolerance film resistors. In the case of carbon composition resistors, we usually add 10% to the purchase tolerance and derate 50%. This approach is too expensive for higher-cost film resistors, which are far more stable. Use of statistically-derived resistor design tolerances can not only result in improved circuit performance, but can save a lot of money.

Data from extensive component testing show that the changes in resistance for each of the individual environmental conditions actually follow normal, or Gaussian distributions. For high-stability film resistors, these changes are very small compared to the purchase tolerance. The data also show that the resistance change resulting from one test is independent of the changes resulting from other tests and that the sequence of the testing has only insignificant effect on the changes recorded.

Therefore, the design tolerance can be calculated as being equal to:

$$(\text{Purchase tolerance})^2 + (\Delta R_1)^2 + (\Delta R_2)^2 + (\Delta R_n)^2 \qquad (\text{Eq. 4-1})$$

Examples are:

Arithmetic Calculation

	Max. ΔR
Purchase Tolerance	±5.0%
Shock	±0.5%
Vibration	±0.5%
Moisture	±1.5%

Arithmetic Calculation Continued

Soldering	±0.5%
Short-time Overload	±0.5%
Load Life	±1.0%
	±9.5%

Root-Mean-Square Calculation

	Max. ΔR	(Max. ΔR)²
Purchase Tolerance	±5.0%	25.00
Shock	±0.5%	0.25
Vibration	±0.5%	0.25
Moisture	±1.5%	2.25
Soldering	±0.5%	0.25
Short-time Overload	±0.5%	0.25
Load Life	±1.0%	1.00
	±9.5%	29.25

$\sqrt{29.25} = 5.41\%$

4.5 HIGH-VOLTAGE RESISTORS

Specially constructed high-voltage resistors (Fig. 4–15) are required in circuits where high continuous working voltages are present, but also wherever the resistor can be subjected to high-voltage pulses and transients (Table 4–11).

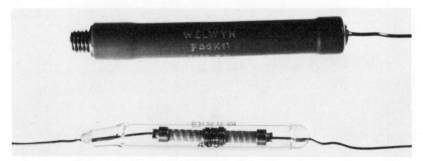

Fig. 4–15 High-voltage and high-value resistors, shown full size. Top: High voltage oxide resistor. Sample shown is 15 MΩ, with 14kV dc working voltage and 50kV pulse voltage. Bottom: Ultra high value Cerox film resistor vacuum sealed in a glass envelope. Sample shown is 500kMΩ, 1/2 watt. *(Courtesy of Dale Electronics, Inc.)*

TABLE 4-11 High-Voltage Resistors

	Cerox® Thick Film Axial, Lead Irradiated Polyolefin Heat-shrunk Sleeve or Glass Capsule	Oxide, Axial Lead in Threaded Insert Each End	Carbon Alloy Axial Leads, Can Withstand 1600°F Flame	Composition Film, Lugs, Terminals or Axial Leads	Ceramic Carbon Composition or Ceramic Silicon Composition, Metalized Ferrules
Power Rating (Temperature)	¼ to ½ w 40°C	½ to 1 w 40°C	½ to 3 w 200°C	1 to 90 w —	15 w to 1000 w 40°C
Derate to 0 watts at	—	—	350°C	—	230°C, 350°C
Voltage WV	0.5, 1.5, 3 kv	4 kv, 14 kv	1 to 15 kv	1.5 to 100 kv	1.5 to 60 kv
Pulse	—	15 kv, 50 kv	—	—	8 to 120 kv
Resistance Range Low	30 MΩ to	2 MΩ to	1 Ω to	10 kΩ to	1 Ω to
High	1000 kMΩ	10,000 MΩ	200 MΩ	1000 MΩ	1 MΩ
Purchase Tolerance	±20%, ±10%, ±5%, ±2%, ±1%	±10%, ±5%, ±2%	±5%, ±2%	±20%, ±10%, ±5%	±20%, ±10%, ±5%

Applications for high-voltage resistors include power supplies, transmitters, X-ray equipment, and electronic precipitators. Power handling and voltage ratings can be increased for some types by immersion in fluids such as mineral oil, silicones, and fluorocarbons; however, resistance may be permanently changed in some types. Consult vendors.

These resistors run larger than ordinary tubular power resistors. For example, a 90-w, 100-kv resistor will be 18.5 in. long, compared to 6.5 in. for a 100-w tubular power wirewound resistor. Mounting must be carefully laid out to avoid unwanted discharge paths, and sharp edges on conductors and connectors must be avoided to minimize corona.

4.6 WHO MAKES WHAT

Table 4–12 lists some of the major resistor manufacturers.

TABLE 4-12   Fixed Resistor Manufacturers

Type	Allen-Bradley	AVX–Aerovox	Carborundum	Clarostat	Corning	Dale	Mallory	Nytronics	Ohmite	Speer	Sprague	Stackpole	TRW/IRC
General-Purpose and Semi-precision													
Carbon Composition	x								x	x		x	x
Wirewound													
Phenolic													x
Ceramic Shell									x		x		
Flameproof						x			x				x
Vitreous Enamel						x			x				
Silicone Ceramic						x			x				
Silicone Coated									x	x			x
Aluminum Heat-Sinked						x		x					
High Power	x			x		x	x		x		x		x
Carbon Film						x			x				x
Metal Film					x	x					x		x
Cermet													x
High-Voltage/High-Resistance			x			x							x
Precision and Ultra-precision													
Metal Oxide Film					x	x							
Metal Film					x	x					x		x
Carbon Film	x					x			x				x
Wirewound						x		x	x				
Wirewound, Aluminum Heat-Sinked						x							x
Wirewound, Bobbin						x							

SUGGESTED READINGS

Bardsley, M., and A. F. Dyson, *Carbon Composition, Cracked Carbon, Metal Film and Metal Oxide Resistors.* The Radio and Electronic Engineer, England, 44, No. 4, April 1974.

Dale Electronics, Inc., *Pulse Handling Capabilities of Wirewound Resistors.* Columbus, Neb., n.d.

Wellard, Charles L., *Resistance and Resistors.* McGraw-Hill Book Co., New York, N. Y., 1960.

CATALOGS

Allen-Bradley, Milwaukee, Wisc.

Corning Glass Works, Corning, N. Y.

Dale Electronics, Inc., Columbus, Neb.

Ohmite Manufacturing Co., Skokie, Ill.

Stackpole Electronics Components, Kane, Pa.

TRW/IRC, TRW Electronics Components, St. Petersburg, Fla.

CHAPTER FIVE

Variable Resistors

Variable resistors used in electronic equipment are known collectively as potentiometers, or pots. More specifically, they can be classed as potentiometers, trimmers, and rheostats (Fig. 5–1). All are three-terminal resistors with one fixed terminal at each end of the resistive element, and a third terminal attached to a movable tap that can be slid along the element, varying the resistance between the tap terminal and the end terminals. The difference between a variable resistor and an adjustable resistor is that the circuit between the resistive element and the movable tap is not opened even momentarily when the tap is moved, as it is in an adjustable resistor.

Potentiometers are designed for frequent and sometimes continuous movement of the tap. They are made in many styles, in commercial, industrial, and military grades, in nonprecision and precision tolerances, and in one-turn and multi-turn styles. Most types can be ganged and have auxiliary switches added; some types can be ganged in pairs with concentric shafts for independent control of the two potentiometers.

Trimmers are designed for only occasional movement of the tap; being used primarily to adjust or trim combinations of circuit component values to correct tolerance buildups, etc. Trimmers are made in one-turn and multi-turn styles, and in commercial, industrial, and military grades. (The term *multi-turn* has different meanings when applied to potentiometers and trimmers.)

The dictionary definition of a rheostat is "a two-terminal variable resistor," but in today's terminology, a rheostat is a wirewound variable resistor of higher wattage, with three terminals. It is gen-

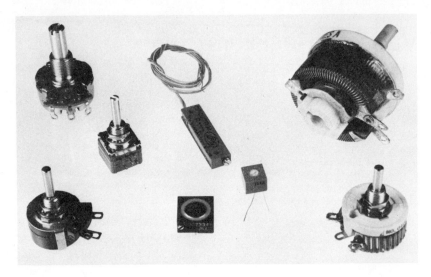

Fig. 5-1 Potentiometers, trimmers, and rheostats. Shown are only a few of the many varieties made. *(Courtesy of Clarostat Mfg. Co., TRW/IRC Potentiometers, Ohmite Mfg. Co.)*

erally used as a two-terminal variable resistor for the purpose of limiting current flow in a circuit.

A word of caution: Although the grade of a potentiometer, trimmer, or rheostat can be surmised by picking it up, looking at it, and reading off the resistance value, the diverse types of resistive elements make it impossible to visually determine the wattage rating merely by the size of the cover or its style.

Carbon composition, resistance wire, conductive plastic and cermet are used as potentiometer resistance elements. Table 5-1 compares the performance of these materials as potentiometer elements.

Power Ratings. The power that can safely be applied to a potentiometer is based on several factors, chief of which is the power that can be dissipated by the resistive element as heat. This power rating is given for each type and variety of potentiometer at some maximum ambient temperature, usually 40°C or 70°C. The potentiometer can be operated at higher ambient temperatures if the applied power is reduced (Fig. 5-2) along a straight line to the maximum operating temperature.

Many potentiometer varieties can be assembled into multiple-section potentiometers (Fig. 5-3). The power rating of each of the

TABLE 5-1 Potentiometer Resistive Element Materials. A Comparison of Best Attainable Performance in Small Single-turn Potentiometer Configurations.

	Carbon Composition	Wirewound	Conductive Plastic	Cermet
Resistance Range	50 Ω to 10 MΩ	1 Ω to 100 kΩ	100 Ω to 5 MΩ	100 Ω to 5 MΩ
Temperature Coefficient	±1200 ppm/°C	±10 ppm/°C†	±100 ppm/°C	±100 ppm/°C
Resolution	Infinite	0.06%	Infinite	Infinite
Linearity	5.00%	0.25%	0.25%	0.25%

†At low resistance values, temperature coefficient can be as high as +3900 ppm/°C.

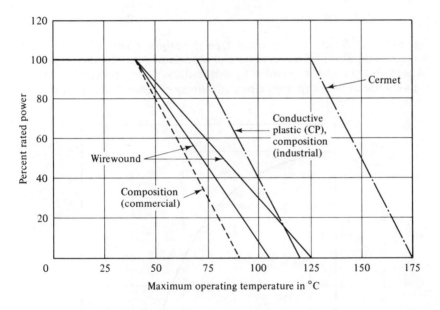

Fig. 5-2 Potentiometer temperature derating.

sections must then be derated, but the method of derating varies from type to type, and vendor catalogs should be consulted. Figure 5-4 shows the required power derating for a typical two-section ganged potentiometer.

Voltage ratings further limit potentiometer power ratings. There is no way to dissipate 2 watts in a 2-watt, 5-megohm potentiometer that has a 500-volts maximum rating. When a potentiometer

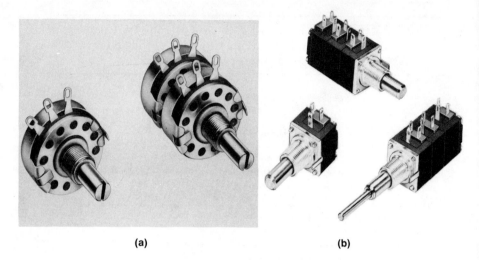

(a)   (b)

Fig. 5-3 **(a)** Single and ganged type J carbon composition potentiometers. **(b)** Modular potentiometers. Models of Allen-Bradley's series
70 Mod Pot® can be assembled by distributors, saving delays and tool-
ing expense in ordering prototypes. *(Courtesy of Allen-Bradley Co.)*

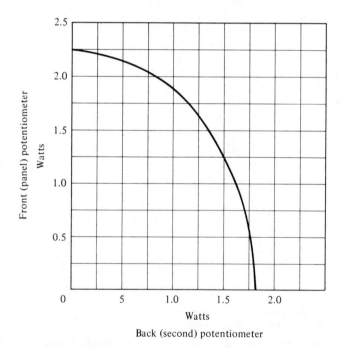

Fig. 5-4 Power derating for type J carbon composition potentiometer.
(Courtesy of Allen-Bradley Co.)

is used as a rheostat, the power rating of the potentiometer must be reduced, because all of the element is not dissipating power. (See Fig. 5–5.) Furthermore, all of the current is now flowing through the wiper contact, and some of these contacts can carry current better than others. General-purpose potentiometers can be used as rheostats within their ratings, precision potentiometers can be used only with caution.

Full-rated power is only obtained with linear taper elements. All others are generally derated 50%.

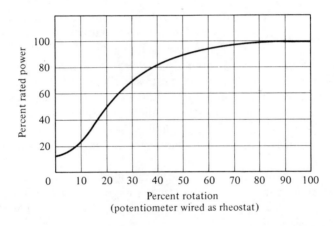

Percent rotation
(potentiometer wired as rheostat)

Fig. 5–5   Potentiometer power derating when the potentiometer is used as a rheostat. *(Courtesy of Allen-Bradley Co.)*

5.1 DEFINITIONS

CONTACT RESISTANCE VARIATION (CRV): In non-wirewound potentiometers, the apparent resistance seen between the wiper and the resistive element when the wiper is moved. Measurement is made under standard energizing current and constant wiper speed. CRV is expressed in ohms or percent of total resistance.

CYCLE: In potentiometer life specification, moving the wiper from one stop to the other and returning. For continuous rotating potentiometers without stops, it is one complete revolution.

END POINT: In precision potentiometers, the shaft positions just beyond the last measurable change in resistance as measured between the wiper and an end terminal with wiper continuity maintained.

END RESISTANCE: In a precision potentiometer, the resistance measured between the wiper terminal and an end terminal with the shaft turned all the way to the corresponding end point.

END RESISTANCE OFFSET: In general-purpose potentiometers, the residual resistance measured between the wiper terminal and an end terminal with the shaft turned all the way to the corresponding mechanical stop. Also called HOPOFF RESISTANCE, END SETTING.

EQUIVALENT NOISE RESISTANCE (ENR): A measure of the electrical noise generated during adjustment of a wirewound potentiometer, stated as an equivalent maximum resistance in series with the wiper.

LINEARITY: The deviation of output versus rotation characteristic of a potentiometer from a straight line. Linearity is defined three ways: Absolute linearity, in which deviation is measured from a straight line drawn between the zero voltage point and the total-applied voltage point; zero-based linearity, in which deviation is measured from the best straight line drawn, starting at zero, through the output curve of the potentiometer being measured; and the loosest definition, independent linearity, in which the reference straight line is drawn through the actual effective travel of the potentiometer, the line being positioned to minimize the maximum deviation.

OUTPUT SMOOTHNESS: In non-wirewound potentiometers, any spurious variation in the output not present in the input, measured with the device connected as a voltage-divider with a dc voltage applied.

PRECISION: The degree of linearity of the output over the resistance range. As in fixed resistors, 1% linearity or better is precision, 3% is non-precision, and the zone between is semi-precision.

RESOLUTION: The smallest incremental change of resistance possible in a wirewound potentiometer, essentially a function of the diameter and spacing of the resistance wire to the element. Carbon composition, conductive plastic, and cermet potentiometers have, for all practical purposes, infinite resolution.

STOP CLUTCH: The device in a potentiometer or trimmer that allows the wiper to idle at the ends of the resistive element without damage while the adjustment shaft continues to be turned.

TEMPERATURE COEFFICIENT: The change in total resistance with a change in temperature of 1°C.

TOTAL RESISTANCE: The resistance in ohms measured between the end terminals of a potentiometer.

USABLE RANGE: The portion of a potentiometer resistive element available for use after subtracting the end resistance, usually expressed as a percent of the total resistance.

WIPER: The moving contact of a potentiometer.

5.2 CARBON COMPOSITION POTENTIOMETERS

Carbon composition potentiometers have been around since the early days of radio, and they are still going strong. They are made in commercial, industrial, and military grades.

Resistive elements are made two ways: Coated film and molded (Fig. 5–6). In the first process, a mixture of carbon, filler, and binder is coated on a ring of insulating material. The surface of the film is processed to minimize abrasion of the film by the sliding contact tap. The contact is brass or phosphor bronze, spring-loaded against the element.

In the molded type, the carbon composition mix is molded into a cavity in a plastic base. In these potentiometers, the moving tap is a carbon brush, giving carbon-to-carbon contact. Molded composition potentiometers are generally totally enclosed and sealed

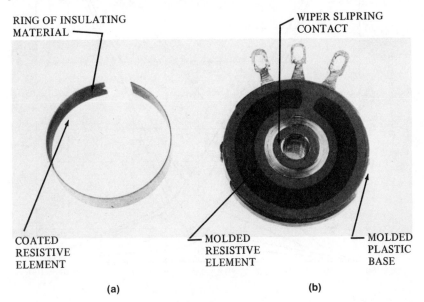

(a)   (b)

Fig. 5–6 Carbon composition potentiometer resistance elements. **(a)** Coated film. **(b)** Molded.

against moisture and other environmental contamination. More rugged than film carbon composition potentiometers, they are widely used in test equipment, computers, servo systems, and other military and industrial applications.

Carbon composition potentiometers are made in linear tapers and in a variety of standard nonlinear tapers, both with right-hand and left-hand rotation (Fig. 5–7). Nonlinear characteristics are produced by combining segments of resistive mixes having different resistances to make up the total length of the film or molded element. (Resistances are varied by juggling the carbon to filler ratio, see Section 4.2.) Although these nonlinear tapers are often called "semi-logarithmic," they are in fact made up of straight-line resistance segments with broad or narrow transition areas where two mixes of different resistivity blend. Tapped elements are also available. Potentiometers can be ganged to make bridged-T, straight-T, bridged-H

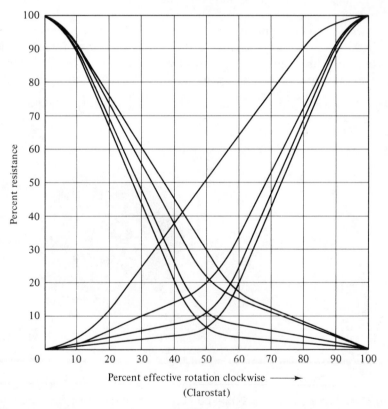

Fig. 5–7   Typical carbon composition potentiometer tapers. *(Courtesy of Clarostat Mfg. Co.)*

and L-pad attenuators. Potentiometer sections, including concentric shaft units and a variety of switches can be combined into single components.

Voltage coefficient, overload, humidity, aging, and temperature characteristics are essentially the same as for molded composition fixed resistors. As inductance and capacitance are both very low, they can be used in many high-frequency applications.

5.3 WIREWOUND POTENTIOMETERS

Small, wirewound resistive elements for commercial, industrial, and military potentiometers are all made in essentially the same way (Fig. 5–8). A single layer of bare resistance wire is space-wound on an insulating core. The resistance of the element can be controlled by varying the diameter of the wire, the cross section of the core, the spacing of the wire turns on the core, the length of the core, or the type of resistance wire used.

FLEXIBLE INSULATOR

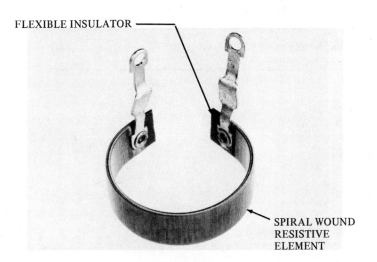

SPIRAL WOUND
RESISTIVE
ELEMENT

Fig. 5–8 Wirewound potentiometer resistive element construction. *(Courtesy of Clarostat Mfg. Co.)*

However, increasing the spacing of the turns on the core or using a larger diameter wire to reduce resistance also decreases resolution. A smaller-diameter wire will increase resistance by both having higher resistance and room for more turns on the core, but fine wire is more fragile and difficult to wind; resolution increase is limited by practical wiper dimensions. Different types of resistance wire (Table 5–2) not only have different resistivities, but the temperature constants are also different, which means that in a given manufacturer's line of potentiometers, not all resistance values will have the same temperature constant, or for that matter, resolution (Figs. 5–9 and 5–10).

TABLE 5-2   Characteristics of Alloys used in Wirewound Potentiometers and Resistors

Alloy	Approximate Composition	Resistivity ohms/circular mil-ft.	Temperature Coefficient ppm/°C
Evanohm*	Ni 75% Cr 20% Al 2.5% Cu 2.5%	800	± 20
Nichrome†	Ni 60% Cr 16% Fe 24%	675	+ 150
Advance†	Ni 43% Cu 57%	294	± 20
Cupron*	Cu 55% Ni 45%	294	± 40
Midohm†	Ni 23% Cu 77%	180	+ 180
No. 90 Alloy*	Ni 12% Cu 88%	90	+ 400
Lohm†	Ni 6% Cu 94%	60	+ 800
No. 30 Alloy*	Ni 2% Cu 98%	30	+1300
Copper	Cu 100%	10.7	+3900

*Trade Mark, Wilbur B. Driver.
†Trade Mark, Driver–Harris.

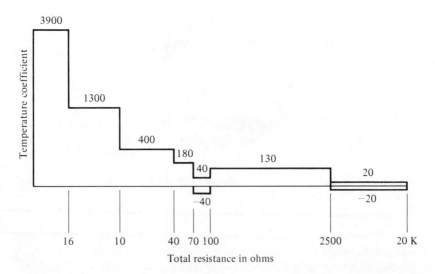

Fig. 5–9 Wirewound potentiometer typical temperature coefficients for different total resistance values. *(Courtesy of Clarostat Mfg. Co.)*

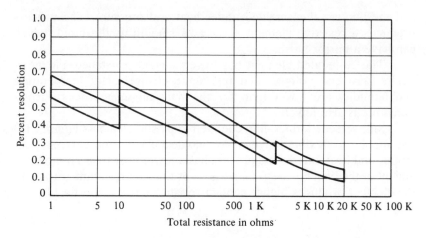

Fig. 5–10 Wirewound potentiometer typical resolution for different total resistance values. *(Courtesy of Clarostat Mfg. Co.)*

Because of their coiled wire construction, wirewound resistors must be viewed as high-resistance inductors. The close spacing of the turns also creates high inter-turn capacitance. Fixed-value wirewound resistors can be non-inductively wound — wirewound potentiometers cannot. They are seldom used above 50 kHz, and they should be used in audio circuits with caution.

5.4 CONDUCTIVE PLASTIC POTENTIOMETERS

Conductive plastic is a carbon–resin mix that can be fabricated into a resistive element three ways: Bulk molding, co-molding, and surface-applied molding (silk-screening, etc.).

Bulk-molded elements are similar in many ways to molded carbon composition elements: The thick element is molded into a cavity in a plastic substrate, and performance characteristics are also similar. They are difficult to manufacture in nonlinear tapers. The resistance change with temperature and humidity is higher than for the other two methods of manufacture, as are contact resistance and end voltage. Bulk elements, however, have extremely long life, typically over 100 million cycles or revolutions with negligible wear and degradation.

In co-molding, the conductive plastic element and a plastic substrate are molded at the same time. A coating of the carbon–resin mix is sprayed on a compacted preform of the substrate material, and the two are then molded under heat and pressure. The process produces a resistive element surface as smooth as the mold, which can have a mirror finish.

The co-molded element has lower contact resistance and practically zero end voltage, and resistance changes with temperature and humidity are lower than for the bulk molding process. Rotational life is also lower.

Silk-screened, thick-film conductive plastic elements (Fig. 5–11) are applied to flexible polyester or molded plastic and ceramic substrates and heat-cured. The process has a lower cost than co-molding, and all of the benefits are retained except that the resultant surface of the element produces higher contact resistance and a shorter life.

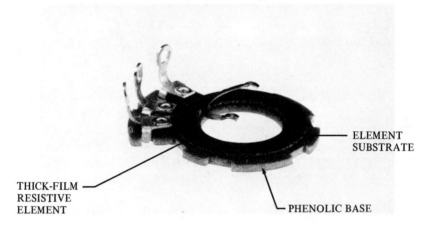

ELEMENT
SUBSTRATE

THICK-FILM
RESISTIVE
ELEMENT

PHENOLIC BASE

Fig. 5–11 Thick-film conductive plastic potentiometer resistive element. *(Courtesy of Clarostat Mfg. Co.)*

5.5 CERMET POTENTIOMETERS

A cermet is a hybrid mixture of a ceramic and a metal. Cermet potentiometer elements are made by screening a mixture of glass and a metal or metal oxide onto a ceramic substrate, and then firing it at a temperature high enough to melt the glass (Fig. 5–12). The result is a hard but unfortunately abrasive film. Cermet potentiometer elements are immune to humidity and have better temperature coefficients than either carbon composition or conductive plastic elements. They don't outgas, because they are completely inorganic and have operating temperatures most often limited by the case material.

KNOB   CERMET RESISTIVE ELEMENT   WIPER

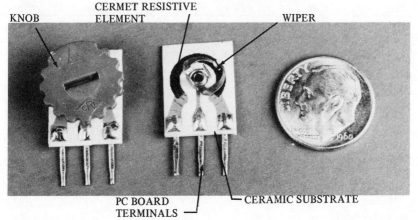

PC BOARD TERMINALS   CERAMIC SUBSTRATE

Fig. 5–12   Cermet potentiometer construction. *(Courtesy of Clarostat Mfg. Co.)*

Table 5–3 compares typical carbon composition, wirewound, conductive-plastic, and cermet general-purpose potentiometers. Several varieties of general purpose potentiometers are shown in Fig. 5–13.

5.6 PRECISION POTENTIOMETERS

Precision potentiometers (Fig. 5–14) are made in a wide variety of single-turn and multi-turn versions. Resistive elements in single-turn potentiometers may be wirewound, conductive plastic, or cermet (see Table 5–4). The advantage of a conductive plastic element over wirewound is infinite resolution; over cermet the advantage is longer life. Elements in multi-turn potentiometers are almost exclusively wirewound.

The case material of industrial precision potentiometers may be either molded plastic or machined metal. Mounting is by threaded bushing. The rotating shaft is usually supported in a single sleeve bearing. Terminals are normally on the backside making ganging impossible, but some models have side terminals and may be ganged together and phased with clamping bands.

Military precision potentiometers have machined metal cases and a standard servo mounting consisting of one or two accurately machined pilot diameters and a clamping ring. Some also have a screw-mounting option. Bearings are ball bearings, on both ends of the shaft. Radial terminals allow ganging and phasing of all units. (See Fig. 5–15.)

TABLE 5-3   General-Purpose Potentiometers, Typical Characteristics

	Wirewound			Carbon Composition	
Typical Example	Clarostat Series 49	Clarostat Series 43	Clarostat Series 58	Allen-Bradley Type J	Allen-Bradley Mod Pot®
Size	$\frac{3}{4}$ in. dia.	$1\frac{1}{8}$ in. dia.	$1\frac{11}{16}$ in. dia.	$1\frac{1}{8}$ in. dia.	$\frac{5}{8}$ in. sq.
Power Rating	$1\frac{1}{2}$ w @ +40°C	2 w @ +40°C	4 w @ +40°C	2 w @ +70°C	1 w @ +70°C
Operating Temperature, Maximum	+125°C	+105°C	+105°C	+120°C	+120°C
Resistance Range	1 Ω to 10 kΩ	1 Ω to 25 kΩ	1 Ω to 100 kΩ	50 Ω to 5 MΩ	50 Ω to 10 MΩ
Standard Tolerances	±5%	±10%	±10%	±20% ±10%	±20% ±10%
Cost: Each, Small Quantity	$5	$1.50	$3	$2	—

Fig. 5–13 General-purpose wirewound, carbon composition, conductive plastic and cermet potentiometers. Except for the wirewound models, the type of element and thus the power rating of any of the potentiometers shown cannot be surmised from exterior inspection. Potentiometers are shown two-thirds size. *(Courtesy of Clarostat Mfg. Co., Ohmite Mfg. Co., TRW/IRC Potentiometers.)*

TABLE 5-3   *Continued*

	Conductive Plastic			Cermet	
Allen-Bradley Type W	Clarostat Series 470	Clarostat Series 388	Allen-Bradley Type CQ	Clarostat Series 389	Allen-Bradley Type CJ
$\frac{1}{2}$ in.	$\frac{15}{16}$ in.	$\frac{1}{2}$ in. sq.	$\frac{1}{2}$ in.	$\frac{1}{2}$ in. sq.	$1\frac{1}{8}$ in.
$\frac{1}{2}$ w @ +70°C	$\frac{1}{2}$ w @ +70°C	$\frac{1}{2}$ w @ +70°C	$\frac{1}{2}$ w @ +125°C	1 w @ +85°C	5 w @ +70°C
+120°C	+120°C	+120°C	+175°C	+150°C	+150°C
100 Ω to 5 MΩ	100 Ω to 5 MΩ	100 Ω to 5 MΩ	100 Ω to 5 MΩ	100 Ω to 5 MΩ	100 Ω to 5 MΩ
±20% ±10%	±20% ±10%	±10%	±10%	±10%	±10%
$4	—	$3	$4	$3	$5

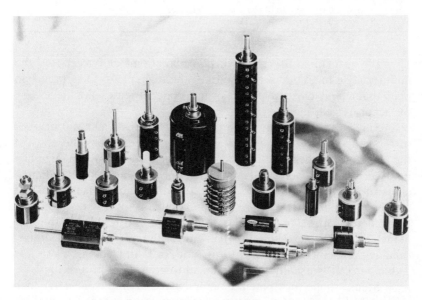

Fig. 5–14 Precision potentiometers. *(Courtesy of Spectrol Electronics Corp.)*

TABLE 5-4 Small Single-Turn Precision Potentiometers†

	Wirewound	Conductive Plastic	Cermet
Power Rating	$2\frac{3}{4}$ w @ +40°C	2 w @ +40°C	5 w @ +40°C
Operating Temperature, Maximum	+125°C	+125°C	+125°C
Resistance Range	5 Ω to 39 kΩ	1 kΩ to 100 kΩ	500 Ω to 2 MΩ
Tolerance Standard	±3%[1]	±10%	±10%
Special	±1%	±5%	±5%
Linearity Standard	±0.5%[2]	±0.5%	±0.5%
Best Practical	±0.25%	±0.25%	±0.25%
Equivalent Noise Resistance	100 Ω	—	—
Output Smoothness	—	0.1%	0.1%
Temperature Coefficient	±800 ppm/°C (5 Ω–20 Ω) ±20 ppm/°C	±500 ppm/°C	±100 ppm/°C
Resolution	0.42% to 0.06%	infinite	infinite
Cost: Each, Small Quantity, Bushing Mount.	$7	$9	$14

†Wirewound, conductive plastic, and cermet resistive elements compared in typical design (Spectrol). All are 1 5/16 in. diameter.
Notes: (1) Above 50 Ω; below 50 Ω ±5% and ±3% tolerances.
      (2) Above 200 Ω; below 200 Ω ±1% and ±0.75%.

If you plan to drive a precision potentiometer by gearing, the military style must be used in spite of the higher cost. The precision pilot diameters, close tolerances, and concentricities are necessary for accurate gear meshing.

Single-turn precision potentiometers have less than 360 degrees of rotary motion to divide the voltage impressed across the ends of the element. As there are practical limits on how fine a wire size can be used in a wirewound resistance element, and limits on the dimensions of a wiper contact face (to be able to make contact with that element), the only way the resolution capability of a wirewound potentiometer can be improved is to increase the length of the element—either by increasing the diameter of the coiled element, by arranging the element in a multi-turn spiral, or both (see Fig. 5–16 and Table 5–5). It is usually more practical to go to a multi-turn unit.

Stock, single-turn potentiometers range in size from a 1/2-inch to 3-inch diameter. Multi-turn potentiometers are made in sizes up to 2 inches with three to ten turns.

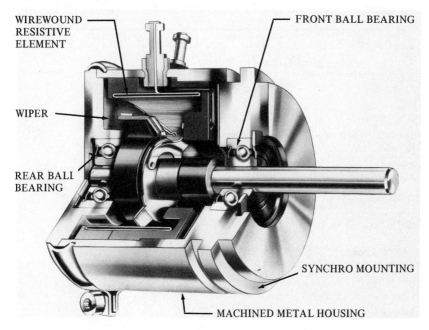

WIREWOUND RESISTIVE ELEMENT

FRONT BALL BEARING

WIPER

REAR BALL BEARING

SYNCHRO MOUNTING

MACHINED METAL HOUSING

Fig. 5–15 Precision potentiometer construction. *(Courtesy of Bunker Ramo Amphenol.)*

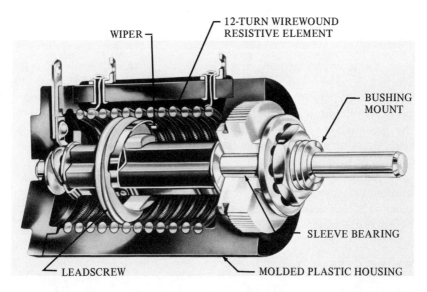

WIPER

12-TURN WIREWOUND RESISTIVE ELEMENT

BUSHING MOUNT

SLEEVE BEARING

LEADSCREW

MOLDED PLASTIC HOUSING

Fig. 5–16 Multi-turn precision potentiometer construction. *(Courtesy of Bunker Ramo Amphenol.)*

TABLE 5-5   Single- and Multi-Turn Wirewound Precision Potentiometers (Typical Characteristics for $\frac{7}{8}$ in. Diameter Varities)

	Single-Turn	Three-Turn	Five-Turn	Ten-Turn
Power Rating	$1\frac{1}{4}$ w @ +70°C	1 w @ +70°C	$1\frac{1}{2}$ w @ +70°C	2 w @ +70°C
Operating Temperature, Maximum	+125°C	+125°C	+125°C	+125°C
Resistance (standard values)	50 Ω to 40 kΩ	50 Ω to 50 kΩ	50 Ω to 100 kΩ	100 Ω to 200 kΩ
Tolerance Standard	±3%	±5%	±5%	±5%
Special	±1%	±1%	±1%	±1%
Linearity	±0.5% to ±1.0%	±0.25%	±0.25%	±0.25%
Equivalent Noise Resistance (ENR)	100 Ω	100 Ω	100 Ω	100 Ω
Shaft Rotation	340°	1080°	1800°	3600°
Temperature Coefficient	±20 ppm/°C	±20 ppm/°C	±20 ppm/°C	±20 ppm/°C
Resolution 100 Ω	0.31%	0.11%	0.75%	0.06%
1000 Ω	0.21%	0.06%	0.04%	0.03%
10 kΩ	0.10%	0.03%	0.03%	0.02%
maximum Ω	0.06%	0.02%	0.01%	0.006%
Rotational Life, revolutions	1,000,000	300,000	500,000	1,000,000
Load Life, hours	900	900	900	900
Cost: Each, Small Quantity Bushing mount.	$ 7	$ 7	$ 7	$ 6
Servo mount.	$11	$13	$13	$12

Turns-counting dials (Fig. 5–17) are used with multi-turn potentiometers, particularly if there is any requirement to be able to return to a setting or to set the potentiometer at a pre-determined point. Dial readouts are either digital or concentric. Most have locking mechanisms to prevent accidental shaft rotation. Not all dials and potentiometers are compatible.

Nonlinear Precision Potentiometers. Functions desired may be defined by formula, empirical data, or plotted curves; computers directly control the nonlinear winding machines. Both variable-pitch windings and step-card techniques are used.

5.7 TRIMMERS

Nobody is perfect. All parts and components have tolerances or are not the value theoretically calculated; nominal values have been

(a)   (b)

Fig. 5–17 Turns counting dials. **(a)** Concentric, model 11. **(b)** Digital, model 15. *(Courtesy of Spectrol Electronics Corp.)*

known to drift. Trimmer potentiometers (Fig. 5–18) are used to correct such deficiencies, to compensate for tolerance buildups and aging, or to vary voltage or resistance to bring a circuit into calibration, or to adjust a circuit to a specific application within a designed range of applications.

A trimmer is not designed for incessant adjustment over its operating life. One hundred cycles would be a lot of activity over the life span; for most types, design life is 200 cycles. If you are planning to adjust a trimmer more often than that, you should consider using a regular potentiometer.

There are two basic trimmer styles: Single-turn and multi-turn. While multi-turn in potentiometers means a long resistance element arranged in a spiral, in trimmers it means that the slider is actuated either by a threaded leadscrew or a worm gear. The result is an improvement in the preciseness of setting by vernier adjustment. Construction of rectangular and square style multi-turn trimmers is shown in Fig. 5–19.

Carbon composition, cermet, and wirewound elements are used in trimmers, with a great variety of package sizes and power ratings (see Tables 5–6, 5–7, and 5–8 on pages 160–62).

Fig. 5–18 Trimmer potentiometers. Shown approximately full size. *(Courtesy of Ohmite Mfg. Co., Clarostat Mfg. Co., TRW/IRC Potentiometers.)*

5.8 RHEOSTATS

A rheostat (Fig. 5–20) is a high wattage — generally higher than four watts — potentiometer. In most applications, rheostats are used as two-terminal variable resistors to limit current flow (Table 5–9). They are also used to a lesser extent as voltage-dividing potentiometers. However, in many such applications, autotransformers and solid-state semiconductor devices are now used because of their greater electrical efficiency.

The size of a high-power rheostat can be formidable, as it is essentially determined by the size of the wire or ribbon used in the element. In most applications, the element conductor size is a function of the current-carrying capacity requirement of the *first turn only*. Once the moving contact is moved to have the second turn in the circuit, the current flow is halved, and so forth every time the resistance in the circuit is doubled. On special order, rheostat windings can be tapered, that is, wound in two to five sections, each succeeding section having wire with a lower current-carrying capacity. The result is a much smaller rheostat, but one having the same wattage rating as a straight-wound model.

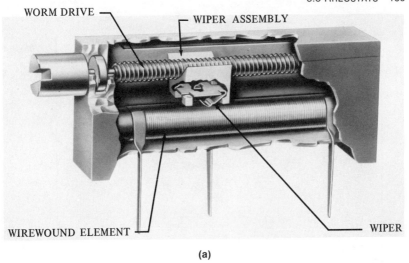

WORM DRIVE

WIPER ASSEMBLY

WIREWOUND ELEMENT

WIPER

(a)

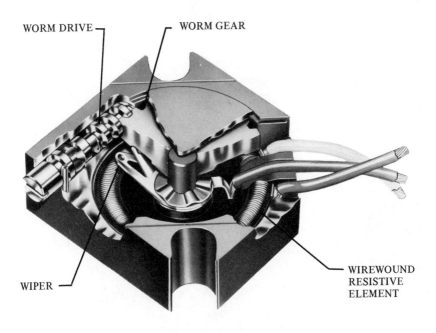

WORM DRIVE

WORM GEAR

WIPER

WIREWOUND
RESISTIVE
ELEMENT

(b)

Fig. 5–19 Multi-turn trimmer potentiometer construction. **(a)** Rectangular. **(b)** Square. *(Courtesy of Bunker Ramo Amphenol)*

TABLE 5-6   Typical Characteristics of Single-Turn Trimmers†

	Carbon Composition	Wirewound	Wirewound	Wirewound	Cermet	Cermet	Cermet	Cermet
Case Style Size	Round ½ in.	Round ½ in.	Cube 5/16 in.	TO–9	Round ½ in.	Round 1 1/32 in. Cube 5/16 in.	Square 3/8 in.	Round ¼ in.
Power Rating	¼ w @ 70°C	1 w @ 70°C	½ w @ 60°C	1 w @ 50°C	¾ w @ 85°C	½ w @ 85°C	½ w @ 85°C	½ w @ 85°C
Operating Temperature, Maximum	+120°C	+125°C	+125°C	+150°C	+150°C	+125°C	+125°C	+150°C
Resistance Range	100 Ω to 5 MΩ	10 Ω to 50 kΩ	50 Ω to 20 kΩ	10 Ω to 50 kΩ	10 Ω to 2 MΩ	10 Ω to 2 MΩ	10 Ω to 2 MΩ	10 Ω to 1 MΩ
Standard Tolerances	±20%	±10%, ±5%	±10%	±5%	±10%	±20%	±10%	±20%
Temperature Coefficient	—	±50 ppm/°C	±70 ppm/°C	±50 ppm/°C	±100 ppm/°C	±100 ppm/°C	±100 ppm/°C	±100 ppm/°C
End Resistance (whichever greater)	15 Ω	1 Ω or 2%	2 Ω to 2%	1 Ω to 2%	5 Ω	2 Ω or 1%	2 Ω	2 Ω
Contact Resistance (CRV)	—	—	—	—	20 Ω or 3%	100 Ω or 10%	1%	3 Ω or 3%
Equivalent Noise Resistance (ENR)	—	100 Ω	100 Ω	100 Ω	—	—	—	—
Rotational Life, cycles	5000	1000	200	500	200	200	200	200
Load Life, hours	1000	1000	1000	1000	1000	1000	1000	1000
Cost: Each, Small Quantity	$3	$4	$2	$4	$2	$2	$4	$2

†All have approximately 300° rotation and can be mounted on PC boards.

TABLE 5-7   Typical Characteristics of Rectangular
Multi-Turn Trimmers†

	Carbon Composition	Wirewound	Cermet
Sizes Made (length)	$1\frac{1}{4}$ in.	$1\frac{1}{4}$ in. $\frac{3}{4}$ in.	$1\frac{1}{4}$ in. $\frac{3}{4}$ in.
Power Rating	$\frac{1}{4}$ w @ +70°C	1 w @ +70°C	$\frac{3}{4}$ w @ +70°C
Operating Temperature, Maximum	+125°C	+150°C	+150°C
Resistance Range	100 Ω to 2.5 MΩ	10 Ω to 50 kΩ	10 Ω to 2 MΩ
Standard Tolerances	±20% ±10%	±10% ±5%	±20% ±10%
End Resistance	0.004% or 20 Ω	2 Ω or 2%	2 Ω to 5 Ω
Temperature Coefficient	–	±100 ppm/°C ±50 ppm/°C	±250 ppm/°C ±100 ppm/°C
Contact Resistance Variation (CRV)	–	–	3 Ω or 3%
Equivalent Noise Resistance (ENR)	–	100 Ω	–
Rotational Life, cycles	500	200	200
Load Life, hours	1000	1000	1000
Cost: Each, Small Quantity	$2–$4	$1–$4	$2–$3

†Number of turns varies 11 to 28; all idle at ends of travel. All available with PC board pins or flexible leads.

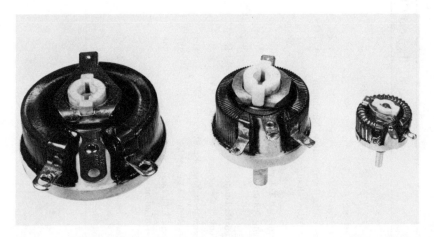

Fig. 5–20   Rheostats; 12 1/2 watt, 25 watt, 50 watt. All shown two-thirds size. *(Courtesy of Ohmite Mfg. Co.)*

TABLE 5-8 Typical Characteristics of Square Multi-Turn Trimmers[†]

	Wirewound			Cermet		
Case Size	$\frac{1}{2}$ in.	$\frac{3}{4}$ in.	$\frac{5}{16}$ in.	$\frac{1}{2}$ in.	$\frac{3}{8}$ in.	$\frac{5}{16}$ in.
Power Rating	1 w @ 70°C	$\frac{3}{4}$ w @ 85°C	$\frac{6}{10}$ w @ 70°C	1 w @ 70°C	$\frac{1}{2}$ w @ 70°C	$\frac{3}{10}$ w @ 70°C
Operating Temperature, Maximum	+150°C	+150°C	+150°C	+150°C	+150°C	+150°C
Resistance Range	10 Ω to 50 kΩ	10 Ω to 50 kΩ	10 Ω to 20 kΩ	10 Ω to 2 MΩ	10 Ω to 1 MΩ	10 Ω to 1 MΩ
Standard Tolerances	±10%, ±5%	±10%, ±5%	±5%	±10%	±10%	±20%, ±10%
Temperature Coefficient	±70 ppm/°C ±50 ppm/°C	±70 ppm/°C ±50 ppm/°C	±50 ppm/°C	±100 ppm/°C	±100 ppm/°C	±100 ppm/°C
End Resistance	1 Ω or 2%	1 Ω or 2%	1 Ω or 2%	2 Ω	5 Ω to 10 Ω	2 Ω
Contact Resistance Variation (CRV)	–	–	–	20 Ω or 3%	20 Ω or 3%	20 Ω or 3%
Equivalent Noise Resistance (ENR)	100 Ω	100 Ω	100 Ω	–	–	–
Rotational Life, cycles	200	200	200	200	200	200
Load Life, hours	1000	1000	1000	1000	1000	1000
Termination PC Board	x	x	x	x	x	x
Lead Wires	x	x		x		
Cost: Each, Small Quantity	$5–$6	$4–$7	$5	$3	$1	$4

[†] Number of turns varies 15 to 25; all idle at ends of travel.

TABLE 5-9   Rheostats, Typical Characteristics

	Ohmite Model C	Ohmite Model E	Ohmite Model H	Ohmite Model J
Size (diameter)	$\frac{1}{2}$ in.	$\frac{7}{8}$ in.	$1\frac{9}{16}$ in.	$2\frac{5}{16}$ in.
Power Rating (40°C)	$7\frac{1}{2}$ w	$12\frac{1}{2}$ w	25 w	50 w
Operating Temperature, Maximum	340°C	340°C	340°C	340°C
Resistance Range (Standard Tolerance ±10%)	10 Ω to 500 Ω	1 Ω to 15,000 Ω	1 Ω to 25,000 Ω	0.5 Ω to 50,000 Ω
Cost: Each, Small Quantity	$19–$22	$7–$8	$7–$8†	$7–$8†

†Prices for high resistance values higher.

5.9 WHO MAKES WHAT

Table 5–10 lists some of the major potentiometer manufacturers. (See page 164.)

SUGGESTED READINGS

Bursky, Dave, "Focus on Pots and Trimmers." *Electronic Design,* August 16, 1975, pp 52–57.

Drew, Bob, "Potentiometers—they're adjusting to changing user needs." *EDN,* April 5, 1975.

Lerch, Jim, "Guidelines for the Selection of Potentiometers." *EDN,* June, 20, 1973.

Stapp, Art, "Potentiometers: Changing to Meet Today's Needs." *EDN,* February 5, 1974.

Wellard, Charles L., *Resistance and Resistors.* McGraw-Hill Book Co., New York, N. Y., 1960.

TABLE 5-10   Variable Resistor Manufacturers

	Allen-Bradley	Bourns	Centralab	Clarostat	Mallory	Ohmite	Spectrol	Stackpole	TRW/IRC
General-Purpose Potentiometers									
Carbon Composition	X			X	X	X		X	X
Wirewound			X	X	X				X
Conductive Plastic				X					
Cermet	X	X							X
Precision Single-Turn Potentiometers									
Wirewound				X			X		X
Conductive Plastic							X		
Cermet							X		
Single-Turn Trimmers									
Carbon Composition	X		X	X		X		X	
Wirewound		X					X		X
Cermet	X	X	X				X		X
Rectangular Multi-Turn Trimmers									
Carbon Composition	X	X							
Wirewound		X		X		X	X		X
Cermet	X	X				X	X		X
Square Multi-Turn Trimmers									
Wirewound		X					X		X
Cermet		X					X		X
Rheostats, All Types					X	X			
Precision Multi-Turn Potentiometers									
Wirewound	X						X		X
Conductive Plastic	X								

CATALOGS

Allen-Bradley Co., Milwaukee, Wisc.

Clarostat Manufacturing Co., Dover, N. H.

Ohmite Manufacturing Co., Skokie, Ill.

Spectrol Electronics Corp., City of Industry, Calif.

TRW/IRC, TRW Electronics Components, St. Petersburg, Fla.

CHAPTER SIX

Nonlinear Resistors

Nonlinear resistors are of two types: Temperature-sensitive *thermistors* and voltage-sensitive *varistors*. In both, current varies, not according to Ohm's law, but as some function of either the body temperature or the applied voltage. Symbols for thermistors and varistors are shown in Fig. 6–1.

Thermistors are resistors with very high temperature coefficients of resistance. The primary function of these devices in a circuit is to change their electrical resistance with a change in body temperature. This characteristic enables simple solutions to many sensing, temperature-compensating, and control problems.

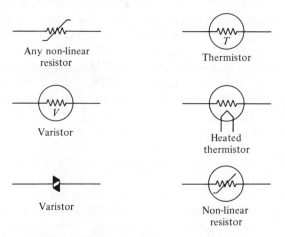

Any non-linear
resistor

Thermistor

Varistor

Varistor

Heated
thermistor

Non-linear
resistor

Fig. 6–1 Thermistor and varistor symbols.

Thermistors are of two basic types: Negative temperature coefficient (NTC) and positive temperature coefficient (PTC). The application of the two types, however, must be considered separately, as the characteristics of one type are not the inverse of the other in polarity, as shown in Fig. 6–2.

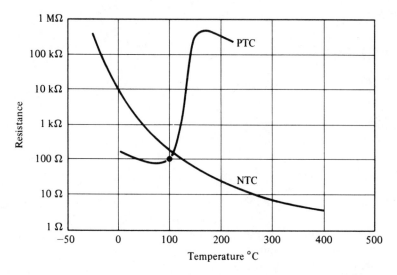

Fig. 6–2 Comparison of NTC and PTC thermistor resistance-temperature characteristics.

An NTC thermistor has a smooth negative-resistance characteristic usable over a wide temperature range; however, a PTC thermistor is usable over a much narrower temperature range, but with greater sensitivity. PTC thermistors can be manufactured to operate over selected temperature bands.

Varistors (Fig. 6–3), also called VDRs (voltage-dependent resistors), are nonlinear resistors with large negative-voltage coefficients. They are generally symmetrical in electrical characteristics.

There are two types of varistors: Silicon carbide, in which the slope of the *E-I* curve changes relatively gradually; and metal oxide, in which the curve changes slope more abruptly. Metal oxide varistors behave in a manner similar to back-to-back zener diodes. When either type varistor is exposed to high-energy voltage transients, the varistor impedance changes from a high standby value to a low conducting value, thus clamping the transient voltage to a safe level. The destructive energy of the incoming high voltage pulse is ab-

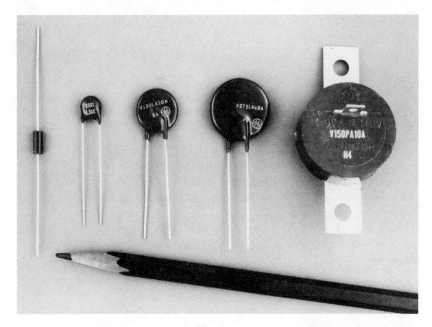

Fig. 6–3   Zinc oxide varistors, General Electric GE-MOV , MA, L, and PA series. *(Courtesy of General Electric Co.)*

sorbed by the varistor, thus protecting voltage-sensitive circuit components.

6.1 THERMISTOR DEFINITIONS

Standardization of terminology has not quite arrived in the thermistor business, but the following definitions are commonly used.

DISSIPATION CONSTANT: The amount of power in milliwatts that will raise the thermistor 1°C. The measurement is normally made at 25°C and in one of three ways: With the thermistor suspended by its leads in still air, immersed in a well-stirred oil bath, or, if a disk or washer, attached to a copper block heat sink.

TIME CONSTANT: The time required in seconds for a thermistor to indicate a change of its own temperature equal to 63% of the difference between a step-function impressed new temperature and its original temperature. The measurement is made under the same mounting conditions as dissipation constant.

BREAKPOINT (PTC THERMISTOR): The point on the rapidly rising portion of the resistance–temperature curve above the Curie point at which the resistance of the PTC thermistor is five times the base resistance.

6.2 NTC THERMISTORS

NTC thermistors are semiconductors of ceramic materials made by sintering various mixtures of oxides of manganese, nickel, cobalt, titanium, copper, iron, and uranium. Electrical characteristics are controlled by varying both the type of oxide used and the physical size and configuration of the thermistor.

Standard configurations for NTC thermistors include beads, glass probes, discs, rods, and washers. Some typical forms are shown in Fig. 6–4.

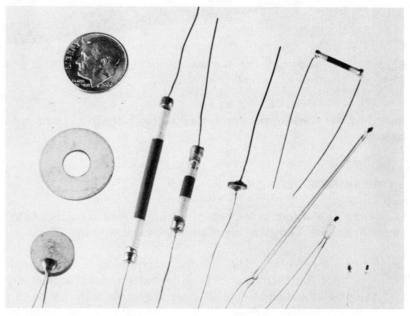

Fig. 6–4 NTC thermistors. Standard configurations, glass beads, rods, discs, washers. *(Courtesy of Fenwal Electronics, Inc.)*

Beads are made by forming small ellipsoids of thermistor material on two fine parallel wires. The material is sintered at high temperature, embedding the leads in the beads. Beads are usually coated with glass for protection for applications to 300°C. Thermistors in the form of beads have small size, fast response, high precision, and stability. Beads range in size from a 0.1-inch to 0.006-inch diameter.

Glass probes are made by sealing beads into the tips of solid glass rods that have a diameter of 0.02 inch to 0.1 inch and a length of 1/4 inch to 2 inches. Glass probes are more easily mounted than beads and are easier to use in liquids.

Discs are made by compressing NTC thermistor material under high pressure in a die. These pieces are sintered and then silvered on both sides. Standard discs range from 0.1 inch to 1 inch in diameter and 0.014 inch to 0.22 inch in thickness. Discs are used when medium power dissipation is needed and when there is sufficient space. Discs have a maximum operating temperature of 150°C.

Washers are similar to discs except that a hole is formed in the center so that the unit can be mounted on a bolt. Washers may be stacked and connected either in series or parallel. Standard washers are 3/4 inch in diameter. Washers are used when high power dissipation is needed.

Rods are die-extruded in several diameters less than 0.188 inch and from 1/4 inch to 2 inches in length. Leads are attached to the ends. The advantage of rods over other types is high resistance with moderate power-handling capacity.

General-purpose NTC thermistors (Table 6–1) must be individually calibrated and are only nominally interchangeable. Purchase tolerances vary up to ±25%, and the slope of the temperature–resistance curve may be different for two thermistors with the same nominal resistance. However, if a thermistor is being wired into a circuit to perform a compensating function requiring one-time circuit adjustment, there is no reason to use a precision thermistor.

Precision NTC thermistors are matched at the factory to a standardized resistance–temperature characteristic, eliminating the need for individual resistance–temperature calibration. Over specified temperature ranges, precision thermistors are electrically interchangeable (Table 6–2).

Thermistor Probe Assemblies. NTC thermistors are packaged in a wide range of standard and custom housings for temperature-sensing application in the fields of medicine, biology, ecology, aerospace, hydrospace, and industry.

Some thermistors are designed for operation at temperatures of up to 1000°C. These devices utilize beads that are spot-welded to two twisted nickel-chrome alloy leads and encapsulated in alumina. Tolerances are broad, and there are minimum and maximum limitations on applied voltage.

TABLE 6-1 General-Purpose NTC Thermistors; Summary Characteristics

		Glass Beads	Glass Beads	Glass Probes	Discs	Washers	Rods	Pellets (Glass Diode Case)
Operating Temperature, Maximum		300°C 400°C	450°C	300°C	150°C	150°C	150°C	300°C
Sizes (inches), Typical		0.005 dia. 0.007 0.010 0.012 0.014	0.043 dia.	0.02 dia. 0.03 0.06 0.07 0.10 0.15	0.05 dia. 0.10 0.20 0.30 0.40 0.50 0.60 0.75 1.00	0.75 dia.	0.05 by 0.5 dia. 0.11 by 0.9 to 2.0 0.17 by 0.9 to 1.8 0.22 by 0.38 to 1.38 0.38 by 1.13	0.07 dia.
Resistance (25°C)		1000 Ω to 9.1 MΩ	30 Ω to 20 MΩ	30 Ω to 20 MΩ	1 Ω to 100 kΩ	1 Ω to 4000 Ω	200 Ω to 2 MΩ	2000 Ω to 1 MΩ
Standard Tolerances		±50% to ±1%	±50% to ±1%	±50% to ±1%	±10% ±5%	±10% ±5%	±10% to ±1%	±10%
Time Constant, (Air)	Min.	0.12 sec	4.0 sec	3 sec	3 sec	150 sec	4 sec	8 sec
	Max.	1.00 sec	5.5 sec	22 sec	300 sec	400 sec	90 sec	–
Dissipation Constant (mw/°C, air)	Min.	0.045	0.35	0.10	1.0	55	2.5	2
	Max.	0.100	0.40	1.70	40	–	125	–
Cost: Each, small quantity		$2–$9	$2–$4	$2–$8	$1–$3	$1–$3	$1–$2	$2–$3

TABLE 6-2   Precision NTC Thermistors†

Form	Epoxy Encapsulated Discs	
Interchangeability Accuracy over Temperature Range	±1% or ±0.2°C	±0.5% or ±0.1°C
Operating Temperature, maximum	−80°C to +150°C	−80°C to +150°C
Size (in.)	0.095 dia.	0.095 dia.
Resistances (250°C) Standard	100 Ω 300 Ω 1000 Ω 2252 Ω 3000 Ω 5000 Ω 10 KΩ 30 KΩ 100 KΩ 300 KΩ 1 MΩ	2252 Ω 3000 Ω 5000 Ω 10 KΩ 30 KΩ
Calibration Temperature Range (typical; varies)	−40°C to +90°C	−40°C to +75°C
Time Constant (Air)	25 sec	25 sec
Dissipation Constant	1 mw/°C	1 mw/°C
Cost: Each, small quantity	$5	$12

†The characteristics of these NTC Thermistors are matched to published curves and are part number interchangeable.

6.3 PTC THERMISTORS

PTC thermistors (Fig. 6–5) are manufactured from doped polycrystalline and semiconducting barium titanate. Powdered barium carbonate, stronthium carbonate, niobium pentoxide, and titanium dioxide are mixed and then calcined at approximately 1200°C. After pulverizing the resultant solid, pellets are pressed and fired in a controlled atmosphere, producing a strong, hard ceramic whose electrical properties are dependent on both the materials used and the thermal and atmospheric conditions maintained during the firing cycle. Electrodes are screened and fired onto the pellets. The adhesion of these electrodes is often stronger than the ceramic itself. Leads are then attached with high-temperature solder. Table 6–3 lists some of the stock types available.

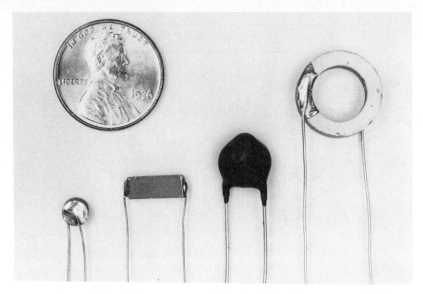

Fig. 6–5   PTC thermistors. *(Courtesy of NL Industries, Inc.)*

TABLE 6-3   General-Purpose PTC Thermistors
Typical Values†

	Disc	Washer	Rod
Operating Temperature	−30°C to +150°C	−30°C to +150°C	−30°C to +150°C
Size (in.)	0.2 dia. 0.3 0.6 0.8	0.60 dia. 0.35 (hole)	0.09 x 0.3
Base Resistance	400 Ω 65 Ω 1000 Ω 18 to 25 Ω	350 Ω 700 Ω	200 Ω to 3200 Ω

†PTC Thermistors are new; data and selection is limited.

The PTC thermistor material is unique in that its resistivity increases abruptly as the temperature is raised above a critical point. This critical temperature, the *break point*, is called the *Curie temperature,* and it can be changed up and down by adjusting the composition of the material, as shown in Fig. 6–6.

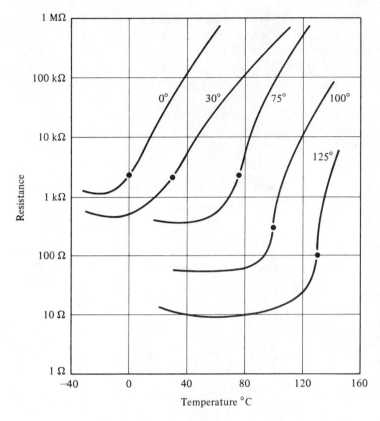

Fig. 6–6   PTC thermistor resistance-temperature characteristics for different breakpoints. *(Courtesy of NL Industries, Inc.)*

6.4 THERMISTOR OPERATION

Both NTC and PTC thermistors can be operated in two modes. In the small-signal or externally heated operational modes, the current is kept low, and the temperature of the thermistor and its resistance are determined by ambient conditions only. In the large signal or self-heated operational mode, the temperature and resistance of both types of thermistors are influenced by power dissipation as well as by changes in the ambient.

Increasing the current through an NTC thermistor beyond the small-signal level will cause enough power to be dissipated within the thermistor to raise its temperature above ambient. When this happens, the resistance begins to go down; the thermistor has negative resistance as the voltage drop across the thermistor goes down with each incremental increase in current. As long as there is enough current applied to the thermistor to hold its temperature sufficiently above ambient (200°C to 300°C), the resistance of the NTC thermistor becomes a function of the internally dissipated power, and the resistance may be 1/1000 of its small signal value (Fig. 6–7).

The resistance of an NTC thermistor is a function of the power dissipated internally under any fixed ambient temperature, providing again that the power level is sufficient to raise the thermistor body temperature considerably above the ambient.

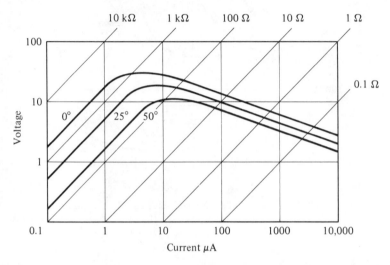

Fig. 6–7 NTC thermistor, E-I steady-state characteristic. *(Courtesy of Fenwal Electronics, Inc.)*

A voltage applied to a resistor and thermistor in series will cause a current to flow that will be a function of total circuit resistance. Assuming that the voltage is high enough, the thermistor will heat, reducing its internal resistance and further increasing the current flow until a steady-state condition exists, determined by the maximum power available in the circuit, as shown in Fig. 6–8. This basic circuit is used for time delay and surge suppression.

Similarly, increasing the current through a PTC thermistor will cause self-heating until a temperature in the region of the steep resistance increase is attained, 170°C, as shown in Figure 6–9. Then, the PTC thermistor will tend to maintain its temperature constant through changes in resistance, with the electrical input power kept equal to the thermal power dissipation. This is the reason that the

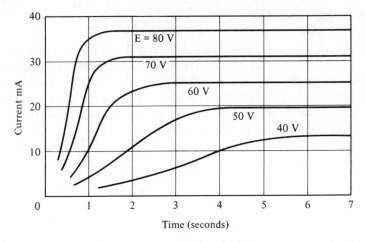

Fig. 6-8 Small NTC thermistor current-time characteristic, voltage applied to thermistor and fixed resistor in series. *(Courtesy of Fenwal Electronics, Inc.)*

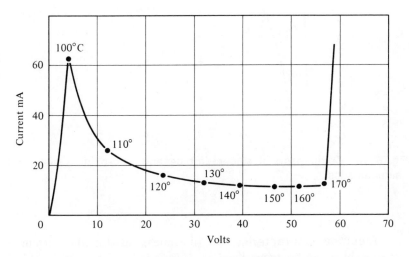

Fig. 6-9 PTC thermistor steady-state E-I characteristic, thermistor curie temperature 125°C. *(Courtesy of Sprague Electric Co.)*

operating temperature of a PTC thermistor changes only little when the supply voltage and ambient temperature are varied in the large signal mode.

6.5 SILICON CARBIDE VARISTORS

These varistors are made from silicon carbide, milled and mixed with a suitable ceramic binder. The material is pressed or extruded to the desired disk, rod, and washer shape and is sintered under controlled atmospheric and temperature conditions to produce a hard ceramic-like material. Small sizes usually have wire leads; the larger washers have surfaces plated to serve as connections (Fig. 6–10).

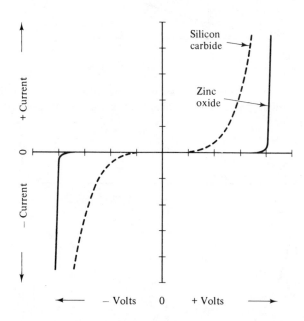

Fig. 6–10 Comparison of silicon-carbide and zinc-oxide varistor E-I characteristics.

Electrical Characteristics. In a linear resistor, the current is $i = e/r$, which can be generalized as

$$\text{current density} = \frac{\text{voltage gradient}}{\text{resistivity}}$$

For a varistor,

$$\text{Current density} = \frac{(\text{voltage gradient})^n}{\text{constant}}$$

The approximate volt–ampere characteristic is:

$$i = \left(\frac{e}{C}\right)^n = Ke^n \qquad\qquad \text{(Eq. 6-1)}$$

Where

$i =$ instantaneous ac or dc

$e =$ instantaneous ac or dc voltage

$C =$ a constant (volts at one ampere)

$K =$ a constant (amperes at one volt)

$n =$ an exponent (1 to 6 depending on processing during manufacture)

The constants c and k depend upon the resistivity, the geometry, and the exponent of the particular varistor under consideration.

The higher the exponent, the more nonlinear the resistance characteristic. To realize the advantages of the high exponent, the applied voltage generally must be kept above one volt. A reduction in voltage gradient is usually accompanied by a reduction in exponent. Depending on geometry and composition of the part, doubling the voltage across a silicon carbide varistor results in the current increasing to as much as 30 times the steady-state value in rise-times ranging from 0.3 to 2.0 microseconds.

Continuous power ratings depend upon permissible temperature rise of the varistors and dissipation of heat. A continuous rating of 0.25 watt per square inch of a varistor surface is usually allowable in still air for separated discs and washers with the plane surfaces vertical; the rating can be increased by radiating fins, forced-air draft, or immersion in oil.

Pulse ratings depend upon the volume of the disc. Assuming no time for radiation, an input of 2000 watt-seconds per cubic inch of varistor will produce a temperature rise of 80°C.

Voltage Surge Protector. Varistors can be used to protect circuits and components from inductive voltage surges by limiting peak discharge voltage to an acceptable level when ac or dc magnetic or inductive circuit current is suddenly interrupted.

If an inductive direct current circuit is opened without a discharge resistor connected across it, theoretically infinite voltage could result, puncturing the coil insulation. When a relay or switch opens the circuit, the circuit is partially protected by an arc following the switch blades, but at the instant the arc breaks, maximum discharge voltage is reached.

A permanently connected linear resistor consumes power under continuous operation, which may be expensive.

The nonlinear resistance characteristics of thyrite material, made first for lightning arrestors, makes it ideal for limiting this inductive kick to a safe value. A varistor used as a discharge resistor connected across a magnetic circuit will protect it whether the circuit is opened by an adjacent or remote switch, or by the occurrence of a fault. Resistance will decrease as the value of induced voltage increases, allowing more current to be drawn through the varistor. The magnetic field energy that would ordinarily force the induced voltage higher will be dissipated in the form of heat by the varistor.

6.6 METAL–OXIDE VARISTORS

The voltage–current characteristic of a metal oxide varistor, like the silicon carbide variety, is Ke^n (Eq. 6–1) except that the exponent, n, is 25 to 50 and up, compared to the range of 2 to 6 for silicon carbide varistors. Metal oxide varistors are packaged as discs, resembling ceramic disc capacitors, axial lead tubulars, or flange-mounted buttons.

At low applied voltages, the metal–oxide varistor is essentially an open circuit. When applied voltage exceeds rated clamping voltage, the device effectively becomes a short circuit, protecting the component that it shunts.

The unit, moreover, requires very little standby power, making it useful for guarding semiconductors. Steady-state power dissipation is typically a fraction of a milliwatt.

The metal–oxide type of varistor has an encapsulated polycrystalline ceramic body, with metal contacts and wire leads. Zinc oxide and bismuth oxide are mixed with other powdered metal additives and then pressed into discs and sintered at a temperature above 1200°C (Fig. 6–11).

As bismuth oxide is molten above 825°C, it assists in the formation of a dense polycrystalline ceramic through liquid-phase sintering. During cooling, a rigid amorphous coating is formed around each zinc oxide grain, producing a microstructure of zinc oxide grains that are isolated from each other by a thin, continuous intergranular phase. This combination is responsible for the nonlinear characteristics.

Table 6–4 describes representative silicon carbide and zinc oxide varistors (pp. 180–81).

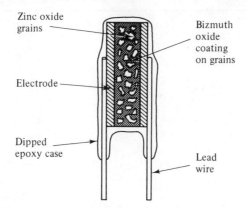

Fig. 6-11   Metal oxide varistor construction.

6.7 WHO MAKES WHAT

Table 6-5 lists some of the major manufacturers of nonlinear resistors.

SUGGESTED READINGS

Capsule Thermistor Course; L-3A. Fenwal Electronics, Framingham, Mass., 1974.

Fabricius, John H., Manfred Kahn, and Vincent T. Guntlow, *Positive Temperature Coefficient Resistors in Temperature-Controlled Heaters;* TP74-8. Sprague Electric Company, North Adams, Mass., 1974.

General Characteristics of Varistors. NL Industries, Inc., Hightstown, N.J., undated.

Golden, F. B., and R. W. Fox, *GE-MOV Varistors Voltage Transient Suppressors;* 200.60. General Electric Company, Auburn, N. Y., 1971.

Harnden, Jr., J. D., F. D. Martzloff, W. G. Morris, and F. B. Golden, "Metal-oxide Varistor: A New Way to Suppress Transients." *Electronics,* October 9, 1972.

Kahn, Manfred, *Characterization and Applications of Barium Titanate PTC Thermistors;* TP71-7. Sprague Electric Company, North Adams, Mass., 1971.

TABLE 6-4 Silicon Carbide and Zinc Oxide Varistors: Range of Values for Representative Styles

Zinc Oxide (General Electric GE-MOV®)

Style	Axial lead tubular (MA series)	Small disc (ZA series)	Button, flange mounted (PA series)
Range of Steady-State			
Applied Voltage DC	26 to 365 vdc	16 to 153 vdc	170 to 750 vdc
AC	20 to 264 vrms	12 to 115 vrms	130 to 575 vrms
Discharge Capacity,			
Joules (watt seconds)	0.1 to 0.7 ws	0.6 to 10.0 ws	10 to 80 ws
Average Power	0.2 w	0.17 to 0.55 w	3 to 10 w
Typical Example	V180M A 3B	V180Z A10	V480PA80C
Steady-State			
Applied Voltage DC	152 vdc	153 vdc	625 vdc
AC	110 vrms	115 vrms	480 vrms
Discharge Capacity	0.3 joules	10.0 joules	80 joules
Peak Pulse,			
Current tp 6 s	20A	1000A	2000A
Max. Average,			
Power Dissipation	0.2 w	0.45 w	10 w
Max. Varistor Voltage			
and Peak Current	320 V (10A)	290 V (10A)	1180 V (10A)
Cost: Each, small quantity	$1–$2	$1–$2	$5–$12

TABLE 6-5   Nonlinear Resistor Manufacturers

	Carborundum	*Fenwal Electronics*	*General Electric*	*NL Industries*	*Texas Instruments*	*Thermometrics*	*Victory Engineering*	*YSI-Sostman*
NTC Thermistors	x	x		x		x	x	x
PTC Thermistors				x	x			
Silicon Carbide Varistors				x				
Zinc Oxide Varistors			x					

TABLE 6-4 *Continued*

Silicon Carbide (NL Industries)

Small disc, $\frac{1}{2}$ in.	Large disc, $1\frac{1}{8}$ in.	Washer, 3 in. dia.	Washer, 6 in. dia.
15 to 150 vdc	30 to 300 vdc	7 to 1500 vdc	150 to 275 vdc
13.5 to 135 vrms	27 to 270 vrms	6.3 to 1350 vrms	135 to 250 vrms
50 ws	275 ws	1100 to 9250 ws	22.500 ws
0.75 w	1.5 w	3 w	10 w
70D–5010	71D–10000	62W–30100	69W–60100
150 vdc	300 vdc	7 vdc	275 vdc
135 vrms	270 vrms	6.30 vrms	250 vrms
50 joules	275 joules	1100 ws	22.500 ws
0.125A	0.250A	10A	12.5A
0.75 w	1.5 w	3 w	10 w
400 V (0.125A)	1000 V (0.25A)	40 V (10A)	1200 V (10A)
$2–$4	$4	$6–$12	$13

Thermistor E-I Curve Manual; L–7. Fenwal Electronics, Framingham, Mass., 1973.

Thermistor Manual; EMC–6. Fenwal Electronics, Framingham, Mass., 1974.

CATALOGS

Carborundum Company, Niagara Falls, N. Y.

Fenwal Electronics, Framingham, Mass.

NL Industries, Inc., Hightstown, N. J. (Varistors, NTC thermistors)

NL Industries, Muskegon, Mich. (PTC Thermistors)

Thermometrics, Inc., Edison, N. J.

Victory Engineering Corp., Springfield, N. J.

YSI-Sostman, Yellow Springs, O.

CHAPTER SEVEN

Relays

A relay is a device that functions as an electrically operated switch. In response to an electrical signal, which may be either steady-state or a pulse, electrical contacts open and/or close in some prearranged fixed combination. The contacts may be in the same circuit as the operating signal or in another circuit or combination of circuits.

Most relays are electromagnetically operated (Fig. 7–1). Current through a coil generates a magnetic field that attracts an armature, which in turn closes or opens the electrical contacts. Operation is in the millisecond range. Although some speed adjustment is possible; a relay in which an appreciable delay can be obtained between energization and actuation is classified as a *time delay relay* and is covered in Chapter 8. Symbols for relays are shown in Fig. 7–2.

Solid-state relays (Fig. 7–3) utilize a semiconductor as a load switching device. Hybrid relays combine electromagnetic relays and semiconductors in various combinations. In one form, an electromagnetic relay is used to provide control and local circuit isolation for a semiconductor switch, in another form, a solid-state amplifier drives a conventional relay providing sensitivity.

The design of a relay, more than that of most components, is a result of highly visible performance parameter trade-offs. A careful understanding of the compromises made in relay construction and adjustment and of the practical design compromises you can make, will not only give you better performance, but will save a lot of money and aggravation.

Relays able to switch high currents are called *contactors*. The dividing line between relay and contactor is fuzzy; considering the

Fig. 7–1   Electromagnetic relays.

relay as an electronic component, relays able to switch more than 25 A have arbitrarily been classed as contactors and excluded from the scope of the book.

Relays are manufactured in great variety. Classification by application is impossible, because each variety of relay is used in many wildly dissimilar applications. In this handbook, relays are sorted by construction features. This scheme also ends up with overlapping classification, but it is easier to work with. The following classes of relays are arranged in order of increasing specialization.

GENERAL-PURPOSE: A relay, usually low-cost, that is adaptable to many applications and is not special in any way.

POWER RELAY: A relay with contacts rated up to 20 A dc or ac.

TELEPHONE-TYPE: A relay once used mainly in telephone systems. The contact springs, interleaved with insulators, are clamped to the frame and actuated by being bent by insulated studs attached to the armature. Contact combinations are often special.

CARD-ACTUATED: A relay in which wire or flat spring contacts pass through holes in a perforated card attached to the relay armature and are actuated by movement of the card.

SENSITIVE: A relay that can operate on a comparatively small amount of power or on a small pulse signal for continuously powered relays.

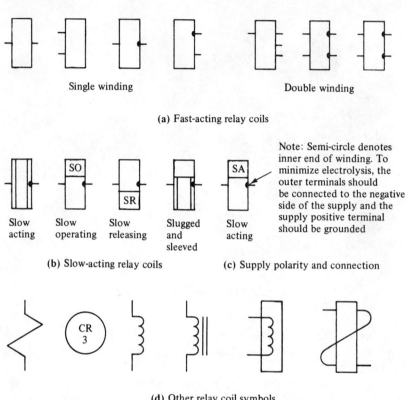

Single winding   Double winding

(a) Fast-acting relay coils

Slow   Slow   Slow   Slugged   Slow
acting   operating   releasing   and   acting
                            sleeved

Note: Semi-circle denotes
inner end of winding. To
minimize electrolysis, the
outer terminals should
be connected to the negative
side of the supply and the
supply positive terminal
should be grounded

(b) Slow-acting relay coils   **(c)** Supply polarity and connection

(d) Other relay coil symbols

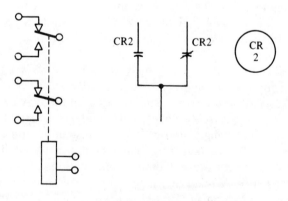

Relay and contacts (form-C contacts shown)

Fig. 7–2 Relay symbols.

Fig. 7–3 Solid-state relay. JDO series relay consists of a thyristor load switch controlled and isolated by an LED optical coupler. Relay is shown full-size. *(Courtesy of Potter & Brumfield Division, AMF Inc.)*

CRYSTAL CAN: A miniature hermetically-sealed relay in a metal case of the same size used for oscillator crystals. The relay is now made in several sizes.

DRY REED: A relay using glass-enclosed magnetic reeds as contact members.

MERCURY-WETTED REED: A reed relay whose contacts are wetted by a film of mercury to provide low-resistance contact and to eliminate contact bounce.

PC BOARD: A relay with pin terminals arranged in patterns for PC board insertion and able to withstand automatic soldering.

STEPPING: A relay in which a single group of contacts is programmed by cams actuated by the relay armature.

COAXIAL: A relay with contacts enclosed in a cylindrical chamber dimensioned to match the impedance characteristic of a particular coaxial cable in order to minimize VSWR and to obtain adequate shielding.

INSTRUMENT: A relay whose movable contact arm is a modified D'Arsonval meter pointer. The stationary contact can be adjusted to any meter dial position.

HYBRID: A relay in which electromechanical and electronic devices are combined to perform switching functions.

SOLID STATE: A relay in which a semiconductor switching device is driven by an amplifier, LED optical coupler, transformer, or gated oscillator.

Most classes of relays can be modified in operational mode or physical configuration to meet the needs of special applications.

7.1 ELECTROMAGNETIC RELAY DEFINITIONS

Although the National Association of Relay Manufacturers (NARM) has been working many years to get relay terms standardized, all manufacturers are not yet using a common language in describing relays. None of the following disagree with the NARM glossary, but the list also includes some additional terms in common usage.

ADJUSTMENT: The modification of tension, shape, position, etc., of relay parts to affect operating characteristics.

ARMATURE: The moving magnetic member of an electromagnetic relay structure.

ARMATURE, BALANCED: A relay armature that rotates about its center of mass and is therefore approximately in equilibrium with respect to both gravitational (static) and acceleration (dynamic) forces.

ARMATURE, CHATTER: The undesired vibration of the armature due to inadequate ac performance. It can also be caused by external shock or vibration.

ARMATURE, HINGE: A pivot provided by a joint, spring, or reed that secures the armature to the remainder of the relay frame or heel-piece.

ARMATURE, OVERTRAVEL, DROPOUT: The portion of the armature travel that occurs between closure of the normally closed contact(s) and the fully released static position of the armature.

ARMATURE, OVERTRAVEL, PICKUP: The portion of the armature travel occurring between closure of the normally open contact(s) and the fully operated static position of the armature.

ARMATURE, REBOUND: (1) The return motion or bounce-back toward the unoperated position after the armature strikes the core on forward motion during pickup, referred to as *armature pickup rebound*. (2) The forward motion or bounce in the direction of the operated position when the armature strikes its backstop on dropout, referred to as *armature dropout rebound.*

ARMATURE, RESIDUAL: The protrusion from the armature that provides the residual gap.

ARMATURE, TRAVEL: The distance that the armature moves in going from its unoperated to its operated position.

BREAK: The opening of closed contacts to interrupt an electrical circuit. (See MAKE.)

BRIDGING: (1) *Normal bridging*—The normal make-before-break action of a make-break or "D" contact combination. In a stepping switch or relay, the commoning together momentarily of two adjacent contacts, by a wiper shaped for that purpose, in the process of moving from one contact to the next. (2) *Abnormal bridging*—The undesired shortening of open contacts caused by a metallic protrusion developed by arcing.

BUFFER: An insulating member that transmits the motion of the armature from one movable contact spring to another in the same pileup.

CARD, ARMATURE: An insulating member used to link movable contact springs to the armature on some types of relay.

COIL: An assembly consisting of one or more windings usually wound over an insulated iron core or on a bobbin or spool, or self-supporting, with terminals, and any other required parts such as a sleeve or slug.

COMB: An insulating member used to position a group of contact springs, as on wire-spring relays. (See CARD, ARMATURE.)

CONTACT, AUXILIARY: Relay contacts used to establish interlocking circuits, to hold a relay in the operated position after the original operating circuit is opened, or to energize a signal to indicate the position of the main contacts.

CONTACT, BOUNCE: The intermittent and undesired opening of closed relay contacts or closing of open contacts caused by one or more of the following: (1) impingement of mating contacts; (2) impact of the armature against the coil core on pickup, or against the backstop on dropout; (3) momentary hesitation, or reversal, of the armature motion during the pickup or dropout stroke.

CONTACT, CHATTER: The undesired vibration of mating contacts during which there may or may not be actual physical contact opening. Even if there is no actual opening, there can be a change in contact resistance. Chatter may result from contact impingement during normal relay operation and release, uncompensated ac operation, or from external shock and vibration.

CONTACT, DOUBLE BREAK: A contact combination in which contacts on a single conductive support simultaneously *open* electrical curcuits connected to two independent contacts, providing two contact air gaps in series when the contact is open.

CONTACT, DOUBLE MAKE: A contact combination in which contacts on a single conductive support simultaneously *close* electrical circuits connected to two independent contacts, providing two contact air gaps in series when the contact is open.

CONTACT, FOLLOW: The distance two compliant contacts travel together after just touching. (See ARMATURE, OVERTRAVEL.)

CONTACT, MOVABLE: The member of a contact combination that is moved directly by the armature.

CONTACT, PERMISSIVE MAKE: A contact combination in which the movable contact spring is pretensioned so that it will close as a result of its own force when unrestrained.

CONTACT RESISTANCE: The electrical resistance across closed contacts as measured at their associated external terminals.

CONTACTS, NORMAL POSITION: The de-energized position of contacts, open or closed, due to spring tension, gravity, or magnetic polarity. The term is also used for the home position of a stepping switch.

CONTACTS: (1) The part of current-carrying members where electrical circuits are opened or closed. (2) The current carrying part of a relay that engages or disengages to open or close electrical circuits. (3) Used to denote a combination or set.

CONTACTS, BIFURCATED: A forked, or branched, contacting member so formed or arranged to provide some degree of independent dual contacting.

CONTACTS, BREAK-BEFORE-MAKE: Double-throw contact sets in which a moving contact opens the old circuit before closing the new circuit.

CONTACTS, MAKE-BEFORE-BREAK: Double-throw contacts so arranged that the moving contact closes a new circuit before opening the old one.

CONTACTS, NONBRIDGING: A term used to describe a contact transfer in which the movable contact leaves one contact before touching the next. The term is normally used with stepping switches or relays, but means the same thing as break-before-make.

CONTACT, STATIONARY: A member of a contact combination that is not moved directly by the actuating system. Also called a *fixed contact*.

CONTACTS, NORMALLY CLOSED: A contact pair which is closed when the armature is in its unoperated position. A "Form B" contact combination.

CONTACTS, NORMALLY OPEN: A contact pair that is open when the armature is in its unoperated position. A "Form A" contact combination.

CONTACTS, SNAP-ACTION: A contact set having two equilibrium positions. The contacts remain in one of the positions with substantially constant contact pressure during the initial motion of the actuating member until a condition is reached at which the stored kinetic energy causes the contacts to abruptly move the alternate position. The contacts usually take the form of one or more precision snap switches actuated by the relay armature.

CONTACT WIPE: The scrubbing action between mating contacts resulting from contact or follow.

DIFFERENTIAL RELAY: A relay, with two or more windings, in which contact position is determined by combinations of power levels in the windings. In polarized types, the polarity of the power as well as the magnitude determines contact position.

ENERGIZATION: The application of power to a coil winding of a relay. Use of the word assumes enough power to operate the relay, unless otherwise stated.

FORM: The operational configuration of a set of relay contacts.

FRAME: The main supporting portion of a relay, which may include parts of the magnetic structure.

LATCHING RELAY: A relay in which contacts are magnetically or mechanically held in the last relay energized position after energization is removed. The contacts can be released to normal position electrically or mechanically.

MAKE: The closing of open contacts to complete an electrical circuit.

OPERATING CHARACTERISTICS: Pickup, nonpickup, hold and dropout, voltage or current (Fig. 7–4).

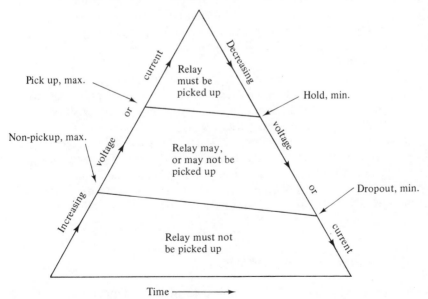

Fig. 7–4 Electromagnetic relay operating characteristics. *(Courtesy of National Association of Relay Manufacturers)*

1. *Dropout value, measured.* The value at which all contacts restore to their unoperated position as the current or voltage on an operated relay is decreased.

2. *Dropout value, specified.* The value at or above which all relay contacts must restore to their unoperated positions as the current or voltage on an operated relay is decreased.

3. *Hold value, specified.* The value which must be reached before any contact change occurs as the current or voltage on an operated relay is decreased.

4. *Nonpickup value, specified.* The value which must be reached before any contact change occurs as the current or voltage on an unoperated relay is increased.

5. *Pickup value, measured.* The value at which all contacts function as the current or voltage on an unoperated relay is increased.

6. *Pickup value, specified.* The value at or below which all contacts must function as the current or voltage on an unoperated relay is increased.

OVERDRIVE: Use of greater than normal coil current (applied voltage), employed to obtain abnormally fast operating time or pulse response.

PILEUP: An assembly of contact springs or combinations fastened one on top of another with insulation between them.

POLE, DOUBLE: A contact arrangement that includes two separate contact combinations — in other words, two single-pole contact assemblies.

RESIDUAL GAP: The length of the air gap in the magnetic circuit, located between the pole face center and nearest point on the armature when the armature is in the energized position. Also called *armature residual gap.*

RESIDUAL SCREW, PIN, PLATE, STUD, OR SHIM: Nonmagnetic part attached to either the armature or the core of a relay that prevents iron-to-iron contact between armature and coil core.

RETURN SPRING: A spring that moves or tends to move the armature to the release (normal) position. Also called *restoring spring.*

7.2 GENERAL CONSIDERATIONS

The parts of a relay (in its simplest form — the clapper relay) are an iron core and its surrounding coil of wire, an iron yoke which provides a low-reluctance path for the magnetic flux — the yoke being shaped so the magnetic circuit can be closed (actually, nearly closed) by a movable piece of iron called the *armature,* and a set of contacts (Fig. 7–5). The armature is hinged to the yoke and held by a spring in such a way that there is an air gap in the magnetic circuit. When an electric current flows in the coil, the armature is attracted to the iron core. Electrical switching contacts are mounted on the armature. When the relay coil is energized, these movable contacts break their connection with one set of fixed contacts and close a connection to previously open contacts. When electric power is removed from the relay coil, a spring returns the armature to its original position. To prevent the armature from remaining stuck to the end of the core because of remanent magnetism (the residual magnetism that remains in the core when no current flows in the coil), a separator made of nonmagnetic material maintains a small "residual" air gap between the armature face and the core.

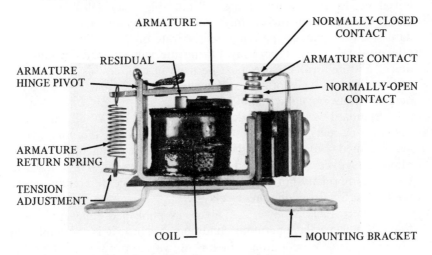

Fig. 7–5 Parts of an electromagnetic relay.

Coils. The relay coil is bobbin-wound with insulated copper wire. The resistance of the coil, and thus the current that can flow in the coil for any energizing voltage, is determined by the gauge of the wire and the length used. The inductance of the coil varies approximately as the square of the number of turns. The coil's magnetomotive force—the force that moves the relay armature—is proportional to the number of turns of wire on the coil times the current (in amperes) flowing through the wire. The more turns put on the bobbin, the less current needed to operate the relay.

Filling the bobbin with a few turns of heavy gauge wire results in a low-resistance, low-voltage relay; whereas many turns of fine wire result in a high-resistance winding that can be operated from a high-voltage supply and that will draw little current to operate the relay. For constant electrical input power and constant magnetomotive force, the current or voltage varies as the square of the coil resistance.

For the same-sized bobbin, a 10,000-Ω coil is more troublesome to make than a 500-Ω coil; it takes a longer piece of wire, more time on the winding machine, and there is more chance of breaking the wire or damaging the insulation during manufacture. So the manufacturer usually charges more for a 110-volt relay than for a 24-volt relay. You might save some money by redesigning a relay driving circuit rather than by using high-voltage relays.

If the number of turns in a relay coil is doubled, only half as much current is needed to operate the relay, everything else being left alone. In other words, a big coil on a big relay will require less power for operation than a tiny coil on a miniature relay. A standard-sized telephone-type relay will operate on 35 milliwatts; a subminiature telephone type relay having the same contact pileup requires 250 milliwatts. Care must be used in making this comparison of two different types of relays, however, because some types are inherently more sensitive than others.

Standard voltages for dc relays are 6, 12, 24, 48, and 110 volts; for ac relays, 6, 12, 24, 48, 120, and 240 volts. Normally, voltage-specified relays are adjusted to pickup at a maximum of 80% of the rated voltage; current-specified relays at rated current or less. Release voltage is normally not controlled. There is a wide differential on dc relays, but a narrow one on ac relays.

Note: Pickup power, as given in published relay specifications, is not a valid operating condition, because it is the minimum power needed for operation, which is no way to run a relay. Operate time will be slower and uncertain, contact resistance will be high, and the contacts can be opened intermittently by shock and vibration. Published pickup power is, however, a valid basis for comparing relay sensitivities.

In ac relays, coil impedance determines whether the relay coil will draw sufficient steady-state current to hold its normally open contacts closed. The ratio of coil impedance to coil resistance is not the same with the armature open as it is with the armature against the pole face separator.

Operate and Release Time. The time interval between the instant the coil circuit of a relay is energized and the closing of the normally open contact is called the *operate time*. The operate time of any coil actuated relay is determined essentially by the resistance and inductance of the relay coil, the adjustment of the relay, effective source voltage, and final steady-state coil current. Relays operating directly from ac operate during the first half-cycle of the energizing signal.

Operate time can be reduced with voltage or current overdrive, or some combination of both. Both are the same if there is no series resistance in the circuit. Series resistance will limit current overdrive without affecting voltage overdrive. Some manufacturers provide curves for calculating operate time as a function of overdrive (Fig. 7–6). You obtain high speed, however, at the cost of increased contact bounce. Polarized latching relays require only a short pulse to operate, in the order of milliseconds, and they can be hit with quite high voltages through protective resistors to limit the coil current.

Release time is the time interval between the instant the coil circuit is opened and the initial opening of the normally open contact. It does not include transfer time. Release time is increased as much as 100 times by a transient-suppressing diode across the relay coil. Capacitor and resistor networks can increase both operate and release time by 10 times. Arcing relay actuating contacts can also increase release time; relay coils energized in parallel will slow each other.

If operate and release times are important parameters in your design, measure them, with the relay installed in your prototype. Some types of relays operate (and release) faster than others. Direct current relays are faster than ac relays; the operate time of some are as short as one millisecond. Low mass armatures, short travel, low eddy current, and biasing help in design.

Adjustment. A strong return coil spring, or many flexed movable contact arms in a telephone relay or a card relay, will slow relay operation but at the same time will provide fast release.

Sensitivity, pickup, nonpickup, hold, dropout, operate time, release time, contact bounce, and sequence of operation (having pairs of contacts opening or closing before or after other pairs) can all be controlled by adjustment within limits. This modification, done by the manufacturer after relay assembly, is called *adjustment*. Telephone relays are also occasionally adjusted in the field.

Adjustment to change one performance parameter upsets other parameters. A close pickup-dropout spread, or an exceptionally wide spread, must be traded-off against high contact forces on the normally open contacts, etc. Environmental shock and vibration characteristics can also be degraded at times.

The operate time and release time of a dc relay can be increased by adding shorted turns to the magnetic circuit, which will

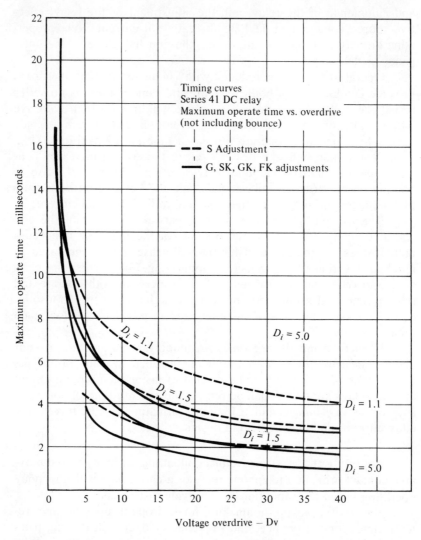

Fig. 7-6 Electromagnetic relay overdrive characteristics. Data shown is for Series 41 sensitive relay. *(Courtesy of Sigma Instruments Inc.)*

produce a counter magnetic force to oppose the buildup and collapse of the magnetic field of the relay coil. The shorted turns are in the magnetic circuit, not in the coil electric circuit: A copper collar or sleeve called a *slug* is mounted on the armature core. To produce a significant time delay, the slug has to be equal in size to or larger than the coil itself, which is why it is most commonly available only on the larger telephone relays. A copper slug on the armature end of the coil produces delay on both pickup and release. On the heel piece end, a slug produces release delay but little pickup delay. A full-length slug produces some pickup delay but a lot of release delay.

Because pickup and release time are affected both by relay adjustments and external circuit conditions, the actual delay obtained is somewhat unpredictable. In any event, delays obtained by slugs are limited to the less-than-1-second range; this is not a way to get precise time-delay functions.

Multiwinding Coils.   Some relays have two or more isolated windings on the coil bobbin or two physically separated bobbins. Differential relays have two or more windings. They are energized in predetermined combinations to produce the magnetic flux necessary to operate the relay. They are made in three types: Conventional, polarized center-off, and polarized-latching.

In the conventional type, the relay operates on the sum of the input currents regardless of net polarity. In a polarized relay, the polarity of the energizing current flowing in the coils, as well as its magnitude, controls the operation of the relay. The windings are sometimes combined with biasing permanent magnets.

A bistable, or double-biased polarized relay, has its armature held in either of two operated positions by permanent magnets (or the flux from a holding coil). When energizing power is removed, the armature remains in the last energized position until flipped to the other position by a signal of the proper polarity and magnitude. This is a *polarized latching relay*.

In a monostable, or single-biased polarized relay, the armature is held in the dropout position by a permanent magnet and will operate only when energized by current of the proper polarity. This type of relay can also be held in the dropout position by flux set up by current flow in a winding or mechanically by a spring.

In a center-off, three-position polarized relay, the armature is restrained mechanically in the neutral, center-off position and thrown to either of two operate positions.

Contacts.   Relay contacts make or break connections in electrical circuits. Current ratings are not the same for dc and ac loads, current ratings generally decrease with increasing open circuit load voltage, and regardless of nominal current rating, contact life varies inversely with increasing load current. None of this applies, however, to contactless solid-state relays, but they have problems of their own, as described in Section 7.15.

Contact Forms. Relay contacts are available in many switching combinations; this feature is the heart of the relay's circuit versatility. These contact variations are called *forms;* standard forms are shown in Fig. 7–7.

Contact Loads. How well contacts perform physically depends on their arrangement, mechanical construction, and materials. Electrically, three factors influence contact performance: The load current magnitude; the open circuit voltage; the characteristics of the load circuit, whether resistive, inductive, capacitive, or incandescent lamp.

Relay load currents are divided by magnitude into four classes: Dry circuit, low-level, intermediate, and power. Dry circuit is not the same as low-level. The dividing lines separating the last three classes are subject to disagreement.

In a dry circuit, there is no current flow when contacts close, nor is there any current flow interruption when contacts open. Once closed, however, the contacts carry current. Contacts must have low interface resistance and be free of organic outgas films and contaminating particles.

Low-level load currents (up to 50 mA) are too small to dissipate contact surface films, to change the contact shape or chemistry at the contact surface, or to soften the contact material.

Intermediate-level load currents (up to 1 A) range higher and dissipate surface films because they raise the contact termperature sufficient to vaporize the film. Arcing in intermediate load contacts erodes the contact surfaces at a generally acceptable slow rate. If the contacts are at the same time fouled with organic outgassing products, carbon will be deposited causing high contact resistance. This is usually more of a problem with hermetically sealed than with open relays. Arcing is the principal limiting factor on the life of the relay. Power load currents (up to 25 A) normally result in arcing and heavy contact erosion. For a resistive load circuit, sizing the contacts is strictly an Ohm's law proposition. With inductive loads, the fast opening of relay contacts causes a sudden high voltage to be generated across the contacts. This voltage—which can be as high as 1000 volts in typical low-voltage relay circuits—is a cause of pitting, welding, material transfer and corrosion of the contacts. Protective circuits such as those shown in Fig. 7–8 can significantly extend contact life, and also decrease the noise generated by the arcing.

In capacitive load circuits, instantaneous current can be very high, whether charging or discharging capacitors. To prevent welding contacts, use series current-limiting resistors.

Incandescent lamp loads, while essentially resistive, create high

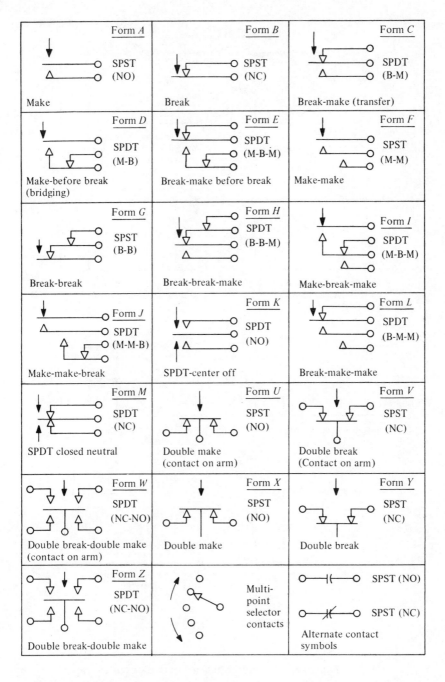

Fig. 7-7 Relay contact forms. The definitions also apply to some types of switches.

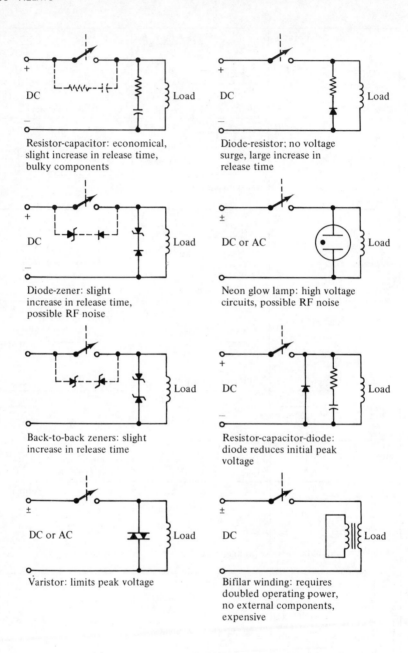

Fig. 7–8 Relay contact protective circuits.

inrush currents because of the low resistance of cold filaments. Contacts should be rated for the inrush current.

Contact wear, or erosion, can cause a variety of problems such as failure to switch, increased contact resistance, decreased insulation dielectric strength, and changes in such relay operating characteristics as pickup and dropout. To assure maximum contact life, some form of arc suppression is recommended in most applications; load characteristics and life requirements largely determine the type and extent of suppression used to assure long life.

In low-level and dry circuit loads, contacts fouled with organic outgassing products can be troublesome, because make-and-break currents are not high enough to dissipate the film.

Contact Materials.   There is no universal relay contact material that is best, good, or even adequate for switching all loads from dry-circuit to 15A (Table 7–1). Platinum contacts should be avoided in low level and dry circuits. Friction heat polymerizes organic materials present, producing a powder that causes high and unstable resistance. Gold, gold alloys, or gold alloys in combination with palladium can eliminate the problem, but gold contacts are prone to sticking or cold welding, and wear faster than silver or palladium contacts.

Gold-plating on silver can reduce contact resistance, but the plating operation must be carefully controlled, and the advantage is short-lived under erosion-causing loads. Gold-plating, however, improves storage life by preventing oxidation.

Under eroding load current operating conditions, palladium is superior to other contact materials. Presence of out-gas fumes will speed erosion, however.

Silver and silver-cadmium contacts are used for high-current loads despite higher contact resistance; the voltage drop across contacts can be as high as 0.20 v, essentially independent of current. Tungsten alloys are used for high-voltage–high-current and inductive load applications.

Relay Life.   Relay mechanical life and electrical life must be considered together. Mechanical life depends on design, wear, material deterioration with time, contamination, and adjustment. From a practical standpoint, electrical life is contact life, and it is a function of contact material and size, load current and voltage, and specific contact operation, i.e., just how the contact pair opens and closes. When long electrical life is required, mechanical life cannot be traded-off against cost.

TABLE 7-1   Contact Materials

	Gold	Platinum	Palladium	Silver	Tungsten
Good Features	Chemically inert Oxide and sulfide free Withstands extreme environmental conditions Long shelf, inactive life	Good resistance to oxidation, corrosion, sticking, and arc erosion Stable contact resistance Withstands extreme environmental conditions	Good resistance to oxidation, corrosion Less expensive than gold or platinum	Best electrical and thermal conductivity. Corrosion resistant	Best arc, corrosion, and erosion resistance No sticking or welding Little wear
Limitations	Expensive Soft Fast-wearing	Soft Fast-wearing	Requires high contact pressure Tendency to weld	Soft Pits and transfers Sulfides form Oxidizes	High contact resistance Oxidizes
Useful Alloys & Purposes	Gold-iridium Gold-palladium Gold-platinum Gold-silver Increase hardness	Platinum-iridium Platinum-osmium Platinum-rhodium Platinum-palladium Better wear resistance	Palladium-copper Palladium-nickel Palladium-platinum Palladium-ruthenium Long life Weld resistance	Silver-copper Silver-cadmium oxide Hardness Higher current rating Long life	Tungsten-carbide Tungsten-silver Tungsten-copper Better conductivity

Enclosures. See Fig. 7–9. Open relays often cost less than relays with dust covers or hermetically-sealed cases. They are also smaller, lighter, and have less trouble from fouled contacts due to outgassing of organic insulating materials. However, covers should be used where there is a dusty or otherwise dirty environment, particularly metal particles floating around (from machine operations, for example), to protect the relay contacts and moving parts from mechanical injury, or service personnel who can't resist attempting mechanical adjustment, and to protect personnel from dangerous voltages. You cannot always take advantage of hindsight and add a cover later to every style relay, so the enclosure decision must be made early.

Dust covers can be metal or plastic, with the trend toward plastic, usually polycarbonate. Most of the time, the dust cover is fitted without a gasket or any other specific sealant; some are held on by a snap-detent or retaining spring clip. Plastic covers can be opaque, transparent for visual observation of relay operation, or in several colors for functional color coding.

Hermetically-sealed enclosures are required for protection from corrosive atmospheric environments, explosion-proofing, long shelf life, and MIL Spec. They should be specified only when absolutely necessary, because they have some problems of their own. Parts must be extremely clean, insulating materials must be selected or aged for minimum outgassing (which can foul contacts), hermetic seals can develop leaks. Before sealing, hermetically-sealed relays are vacuum-baked to remove all air, moisture, and outgas from the enclosure, then are filled with an inert gas, usually dry nitrogen.

Mounting and Terminals. Most varieties of relays are available with a selection of mounting and terminal arrangements. Simplest is a single-stud mount and solder tabs, used on many low-cost general-purpose relays. Telephone relays are usually mounted with screws into the heel piece with solder or wire-wrap terminals. Plug-in relays, combining mount and terminal, are made in a wide variety, including tube sockets (miniature, octal, and octal-derived) and rectangular relay sockets. Plug-in relays should always be enclosed varieties so you have something to safely get a hold of when handling.

Military specification relays are available with clamps, brackets, or stud mounting. Most varieties of relays can be had with pins for PC board mounting. Other terminals available include screw-type, threaded stud, quick-connect, and dip-solder.

Fig. 7–9 Relay enclosures. Most relay types can be obtained with a variety of enclosures. Shown is Sigma's series 41 sensitive relay. From top: Hermetically-sealed, octal plug; dust cover, 5 pin tube plug; open, PC terminals; clear polycarbonate and aluminum dust covers, octal plug; open, screw mounting solder tabs; open, stud mounting, armature common with studs, solder terminals. Relays are shown one-half size. *(Courtesy of Sigma Instruments Inc.)*

Marking. Enclosed relays are marked with circuit diagrams identifying terminal connections. With open relays, you generally have to rely on your eyesight to see what connects to what.

7.3 GENERAL-PURPOSE RELAYS

General-purpose relays (Fig. 7–10) are customarily negatively defined as relays that are not designed for some particular application. Some vendors stretch even this definition to include "general-purpose" telephone type relays, or "general-purpose" reed relays, based on the idea that these "general-purpose" relays are available for delivery either from manufacturer or distributor stock. Confusion can be avoided if a tighter definition is used.

Fig. 7–10 General-purpose relays. Potter & Brumfield GA, KL, KM and KHP series relays, Sigma 68 series relay. All are shown one-half size. *(Courtesy of Potter & Brumfield Division of AMF Inc., Sigma Instruments Inc.)*

A general-purpose relay can be defined as a *clapper* relay suitable for a wide variety of not-too-critical applications—applications where close control of operate time, release time, pickup, dropout, etc., is not important, dry-circuit switching is not being done, and where the relay does not have to operate in high shock and vibration environments. This leaves us with the job of defining a clapper relay, which is a slight problem, because every relay with a moving armature can be thought of as a clapper relay. For our purposes, a clapper relay is a relay with a hinged *straight* (not L-shaped) armature and with the contacts mounted on and a part of the armature (see Fig. 7–5).

Within this restricted definition there is a great variety of general-purpose relays, all comparatively inexpensive. Since they are well-stocked by distributors and manufacturers, they are readily available. These relays can meet most requirements for switching light- to medium-power loads. Typical applications include controlling small motors, solenoids, and other relays in vending machines, photocopiers, industrial process control, and automation equipment.

General-purpose relays can be purchased open, with metal or plastic dust covers, and in hermetically-sealed cases. Plastic dust covers (usually high-strength polycarbonate) can be opaque, transparent so relay operation can be observed, or in different colors for functional color coding. Some types are listed by Underwriters Laboratories, Inc. (UL) and Canadian Standards Association (CSA) for component use.

While the electrical life under load for general-purpose relays is typically 100,000 operations, some varieties have electrical life in excess of 1 million operations.

General-purpose relays can be classed by size: Miniature, small, and standard. Standard-sized general-purpose relays can be further divided into two externally dissimilar groups by the type of plug-in socket used (Table 7–2).

Miniature general-purpose relays—those having a mounted size smaller than one cubic inch—have contacts from SPDT to 3PDT with current ratings of 1 or 2 A. Some varieties are dc only, and few varieties are made in a full range of coil voltages. Miniature general-purpose relays are commonly used in small business machines, tape decks, automatic toys, and alarm systems.

Small general-purpose relays, between 1 and 2 cubic inches in mounted size, have contacts from SPST–NO to 4PDT with current ratings of 2 A and 3 A. Most small general-purpose relays are made either in plastic dust cover or open versions.

Standard-size and general-purpose relays have contacts up to 4PDT with current ratings of 3 A, 5 A, and 10 A. A few have ratings of 15 A. The size of the relays ranges from 2 to 12 cubic inches.

Table 7-2 General-Purpose Relays — Summary Characteristics

	Standard	*Small*	*Miniature*
Standard Enclosures	Open / Dust cover / Hermetically sealed	Open / Dust cover	Open / Dust cover
Plug Type	Octal	Rectangular	None
Weight, Typical — Open	1.7 oz	1.6 oz	0.6 oz
Dust cover	2.3 oz	2.9 oz	—
Hermetically sealed	3.0 oz	—	—
Contacts, Standard	SPST to 4PDT	SPST to 4PDT	SPDT, 3PDT
Ratings, Standard	3 A, 5 A, 10 A	2 A, 3 A	1 A, 2 A
Coil Power	ac, dc	ac, dc	most dc, some ac & dc
Coil Voltages	6, 12, 24, 48, 110 vdc / 6, 12, 24, 48, 120, 240 vac	6, 12, 24, 48, 110 vdc / 6, 12, 24, 48, 120 vac	6, 12, 24, 48 vdc / 120 vac
Rated Coil Power (dc)	1200 mw	900 mw	400 to 1000 mw
Operate Time, Typical	15 to 30 ms	12–20 ms	30 ms
Release Time, Typical	6 to 10 ms	6–10 ms	30 ms
Life, Mechanical, Typical		10 million ops.	
Life, Electrical, Typical		100 thousand ops.	
Operating Temperature, Typical	−45°C to +70°C	−45°C to +70°C	−45°C to +55°C
Cost: Each, small quantity — Open	$4–$5	$4–$8	$1–$5
Dust cover	$7–$10	$9–$11	—
Hermetically sealed	—	—	—

7.4 POWER RELAYS

Power relays (Fig. 7–11) are heavy-duty clapper relays capable of multipole switching of resistive leads of 15 A to 25 A per pole at either 28 vdc or up to 230 vac. They are designed for use in industrial equipment for controlling fractional horsepower motors, solenoids, and heating elements or to energize a high-power contactor.

Fig. 7–11 Power relay. Heavy duty PR series relays for industrial control applications have silver contacts rated at 25 A. Relay carries U/L and CSA listing. Shown half-size; relay is also available with dust cover. *(Courtesy of Potter & Brumfield Division, of AMF Inc., Sigma Instruments Inc.)*

These rugged relays typically have load-life ratings of 100,000 operations, with mechanical life in the millions of operations. Their ability to handle large inrush currents without contact welding is mainly due to contact wiping action. Contacts are of silver or silver alloy, in large buttons. Some relays use snap–switch contacts; others use double contacts that provide two contact air gaps in series on a single pole when the contacts are opened. Open-style relays may be flange- or stud-mounted or mounted on heavy, molded-plastic bases. Power relays are also made in dust cover versions with plugs (both octal and rectangular) or with solder or quick contact tabs, or screw terminals, or in hermetically-sealed metal cases (with plugs or hook solder terminals). Some types are UL- and CSA-listed.

Some types use a magnetic blowout, which is a permanent magnet placed close to the contacts to quench arcs developed across opening contacts in high-voltage dc switching. Table 7–3 shows the characteristics of some typical power relays.

TABLE 7-3 Typical Power Relays

	Magnecraft Class 89 CQ	Sigma Series 55	Potter & Brumfield KR Series	Magnecraft Class 99 DB
Enclosure	Dust cover	Open	Open or dust cover	Hermetically sealed
Connection (Contact)	Quick-connect tabs	Screw terminals	Octal-plug solder	Screw terminals
Mounting	Flange	Flange	Stud or plug-in	Molded plastic base
Contacts	DPST DPDT	DPDT 3PDT, 4PDT	SPST (double)	SPDT (double) with magnetic blowout
Contact Rating (Resistive)	20 A @ 28 vdc or 120 vac	10 A @ 28 vdc 15 A @ 230 vac	20 A @ 28 vdc 20 A @ 120 vac	20 A @ 110 vdc 8 A @ 220 vdc 2 A @ 500 vdc
Operate Time	—	25–30 ms	15 ms	—
Release Time	—	20 ms	10 ms	—
Coil Power, Rated	3 w dc; 7 vA ac	2 w dc; 6–8 vA ac	1.2 w dc; 2.0 vA ac	2 w dc; 10 vA ac
Coil Voltages	6, 12, 24, 110 vdc 6, 12, 24, 120, 240 vac	12, 24, 110 vdc 115, 230 vac	6, 12, 24, 48, 110 vdc 6, 12, 24, 120, 240 vac	6, 12, 24, 110, 220 vdc 24, 115, 230 vac

7.5 TELEPHONE RELAYS

The telephone relay is the most reliable, time-proven, and versatile design in use today. It is made in a variety of sizes from standard to subminiature (see Fig. 7–12 and Table 7–4).

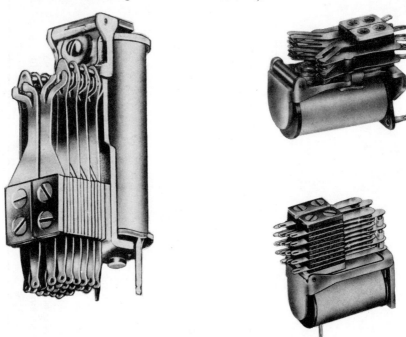

Fig. 7–12 Telephone relays. Left, Standard-size Automatic Electric Class B; right, Potter & Brumfield LS and MH Class relays. All are shown half-size. *(Courtesy of GTE Automatic Electric and Potter & Brumfield Division of AMF Inc.)*

Contacts are available with a choice of materials suitable for load ratings from dry circuit to 50 A. There are contact configurations with low dielectric loss and capacitance for radio–frequency and video-switching, and there are wide-space contacts for switching high voltage ac. For dc operation, coils can be single, concentric, or two-section double-wound. DC relays can be equipped with heel-end or armature-end copper slugs or copper core-sleeves for slow operation. For ac operation, laminated cores are usually used, but dc coils with integral semiconductor diode rectifiers can also be provided. Operating characteristics—pickup, non-pickup, hold, and dropout—can be altered by factory adjustment or by adjustment in the field.

TABLE 7-4   Telephone Type Relays*

	Standard	Medium	Small	Miniature	Subminiature
Height, unenclosed	$4\frac{1}{8}$ in.	$2\frac{1}{2}$ in.	$2\frac{1}{16}$ in.	$1\frac{1}{2}$ in.	$1\frac{1}{4}$ in.
Weight	8 to 12 oz.	6 oz.	3 oz.	2.5 oz.	2 oz.
Pickup power, min. per movable arm	35 mw	60 mw	100 mw	150 mw	250 mw
Pickup time, typical	10 ms	10 ms	10 ms	12 ms	6 ms
Dropout time, typical	10 ms	15 ms	13 ms	7 ms	3 ms
High-frequency (RF & video)			x		x
AC operation	x		x		
Springs per pileup (max.)	20	12	15	10	6
Max. poles (Form C)	12PDT	8PDT	10PDT	6PDT	4PDT
Contact Materials					
Gold Alloy	x	x	x	x	x
Palladium	x	x	x	x	x
Palladium-Silver	STD	x	STD	x	No
Silver Cadmium Oxide	x	x	x	x	x
Silver Tungsten Alloy	x	x	x	x	No
Standard Variations					
Bifurcated	STD	x	x	x	No
Snap-action	x	x	x	No	No
Hermetically Sealed	x	x	x	x	x
Magnetic Latching	x	No	x	No	No
Mechanical Latching	No	No	No	x	No
Mixed Contact Ratings	x	x	No	No	No
Slow Operation	x	x	x	No	No
Sensitive Adjust or Version	x	x	No	No	No

*Width dimensions depend on pileup of contacts. All sizes are available unenclosed for operation inside equipment and closed cabinets with metal and plastic dust covers and hermetically sealed enclosures. Coils are provided in all standard voltage ranges; in the larger sizes, the coil resistance for each nominal voltage depends on the number of contact poles. All have either one or two pileups. Terminal selection includes solder, taper tab, solderless wrap, PC (except standard size), octal type sockets, special relay sockets, and hook terminals in glass headers.

All telephone relays have in common an L-shaped frame or heel-piece that parallels the coil and an L-shaped armature pivoting on the end of the frame. The contact pileups (either one or two) consist of flat cantilevered springs, with stationary and movable contacts interleaved and separated by insulating spacers, firmly attached to the frame with screws. The armature actuates the contacts by means of insulated rods called *buffers,* with additional buffers between moving contacts (Fig. 7–13).

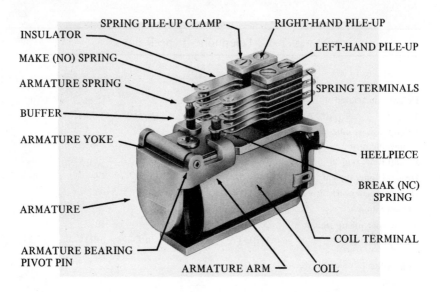

Fig. 7–13 Telephone relay construction. Potter & Brumfield MF Series relay. Shown full size. *(Courtesy of Potter & Brumfield Division of AMF Inc.)*

It is when you get into the area of nonstandard options that the full versatility of the telephone relay is realized. If you don't see what you want in the catalog, ask.

In contacts, there are many combinations of forms, rating, and contact material which can be obtained on a single small, medium, or large telephone relay, within limitations of the maximum total number of contact springs that the coil can actuate. For example, low-level audio or video and high-level power could be switched on the same relay, although you might not want to. Special switching sequences can be provided; for example, have one pair of contacts make before another. This can be done by having different spacing between contact blades or by bending the contact blades.

Standard coils are essentially unaffected by humidity and temperatures up to 105°C. High-temperature coils, impregnated coils, and coils with special resistance and ampere–turn values can be obtained on special order.

7.6 CARD-ACTUATED RELAYS

In this type of relay, the movable contact arms protrude through holes in a rigid plastic card attached to the armature. When the relay is operated, the card moves, either by flexing the movable contact arms to make contact with fixed arms or by allowing prestressed contact arms to make with fixed contact arms. Card-actuated relays are of two general types—*industrial* and *wire spring*—as shown in Table 7–5 on page 212.

Industrial card-actuated relays (Fig. 7–14) are similar in design to telephone-type relays except for the card actuation. Contacts range from DPDT to 8PDT; all standard dc and 115 vac coil voltages are generally available. Sensitive versions are also made. In size, industrial relays are equivalent to subminiature telephone type relays but considerably cheaper because of lower manufacturing costs.

Fig. 7–14 Industrial card-actuated relay. Sigma series 70 shown full size with cover removed. *(Courtesy Sigma Instruments Inc.)*

In a telephone relay, each moving contact arm in a pile-up is pushed by a knob, or "buffer," on the moving arm below it. Accumulating buffer wear results in loss of contact pressure, with the topmost contact set losing pressure the fastest. The design of the card-actuated relay bypasses this problem: Since each moving contact arm protrudes through a separate hole in the actuating card, there is no difference in wear, and there is no build-up of manufacturing tolerances as with buffers.

In wire-spring relays (Fig. 7–15), the contact arms are thin round wires rather than flat springs. A single wire-spring relay can transfer as many as 51 circuits at the same time, replacing four or more conventional relays.

In construction, these relays can differ considerably from tele-

TABLE 7-5 Card-Actuated Relays—Typical Characteristics

	Industrial Relays		Wire-Spring Relays
Examples	Magnecraft Class 67 Sigma Series 70	Automatic Electric HQA	Automatic Electric WQA
Enclosure	Plastic dust cover	Plastic dust cover	Plastic dust cover, or hermetically sealed
Weight	1.0 to 1.4 oz.	1.6 oz.	2¼ lbs.
Mounting	Chassis, or socket or PC board	Socket or PC board	Baseplate and bushings, or studs
Terminals	Solder tab or PC pin	PC pin	Wire wrap, or wire headers
Contact Configuration	Bifurcated or plain, crossbar, button	Bifurcated	Single, coined contacts
Contact Rating dc	1, 2, 3, 5 A	2 A	3 A
Contact Rating ac	1, 2, 3, 5 A	—	0.5 A
Coil Voltages	6, 12, 24, 48, 115 vdc	4, 6, 9, 12, 14, 18, 24, 30, 48, 60, 110 vdc	6, 12, 24, 48, 60, 110, 220 vdc
Contacts, Standard	DPDT 4PDT 6PDT 8PDT	6PDT	17PDT, 34PDT, 51PDT
Rated Coil Power (dc)	1 w 1 w 1.5 w 2 w	2.4 w	4 to 6 w
Pickup Power (min.)	580 mw 580 mw 1000 mw 1200 mw	350 mw	—
Operate Time (typ.)	12 ms 14 ms 16 ms 18 ms	5 to 25 ms†	30 to 50 ms
Release Time (typ.)	8 ms 8 ms 8 ms 8 ms	25 to 30 ms	3 to 5 ms
Life, Mechanical (typ.)	10 million ops.	100 million ops.	1 billion ops.
Life, Electrical, Rated Load (typ.)	200 thousand to 1 million ops.	500 thousand ops.	300 thousand (3 A) 10 million (1 A)
Operating Ambient Temperature (typ.)	−55°C to +71°C	−55°C to +85°C	−55°C to +85°C
Vibration	5 g	20 g (non op.)	5 g (non op.)
Shock	25	50 g (non op.)	20 g (non op.)

†With overdrive.

Fig. 7-15   Card-actuated wire spring relay. Automatic Electric Class W relay (shown full-size) has 51 form C contacts, is available in quick-acting and remanent magnetic-latching models. *(Courtesy of GTE Automatic Electric)*

phone relays. In one version, the coil is mounted on the center leg of an E-shaped core; this configuration, with a large, flat armature, provides an exceptionally efficient magnetic structure. And, unlike most relay form C contact sets that have a movable contact arm between fixed normally-open and normally-closed contacts, the common contact is fixed. All of the contact wires pass through holes in a rigid perforated actuating card; both moving contacts are bent such that they would both make with the fixed common contact if the holes in the actuating card did not hold them open — one open in the relay operate position and the other open in the release position (Fig. 7-16). Movement of the card by the relay armature permits either contact to make; this operation is called *permissive make*.

Applications include circuit-testing apparatus and computer peripheral equipment. The wire-spring relay provides high-contact densities and long life, with low insulation leakage, high switching circuit voltage capabilities, and low intercontact capacitance.

Contact current is limited to 3 A noninductive with the standard palladium–silver contacts. Gold overlay contacts are mandatory for low-level circuits.

The wire-spring relay is made in two versions: A basic quick-acting relay for ordinary service and a magnetically latched bistable relay that uses remanent (residual) magnetism of the coil core for latching.

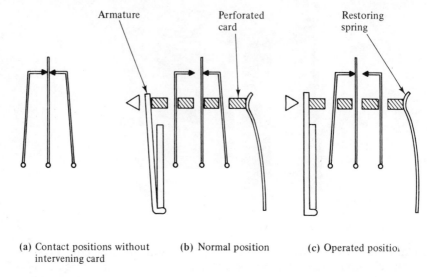

(a) Contact positions without   (b) Normal position   (c) Operated positio̱
intervening card

Fig. 7–16 Card-actuated wire spring relay operation.

7.7 SENSITIVE RELAYS

A sensitive relay (Fig. 7–17) is a relay that can operate on a comparatively small amount of power or, in the case of some relay types that are powered continuously, on a small signal, usually a pulse. Sensitivity is a comparative parameter; some types of relays are inherently more sensitive than others, and within a type, obtainable sensitivity varies with style, construction, number of contact circuits, and size. Some types of relays, notably telephone type relays, can be adjusted to improve sensitivity, while others cannot.

Conventional electromagnetically actuated relays can be made sensitive by increasing the turns on the coil resulting in a stronger magnetic flux (assuming the same coil current), and by increasing the magnetic efficiency of the iron core and magnetic structure both by configuration and choice of materials. Adjustment to reduce the magnetic force necessary to operate the contacts can be made to the various parts of the relay by use of low-friction instrument bearings on the armature, weaker return spring, smaller contact air gaps, etc. Sensitivity obtained by such adjustments is usually obtained at a cost of reduced ability to operate under shock and vibration. However, a well-designed relay adjusted for sensitivity will have no trouble surviving 100 g shocks, although it won't operate reliably

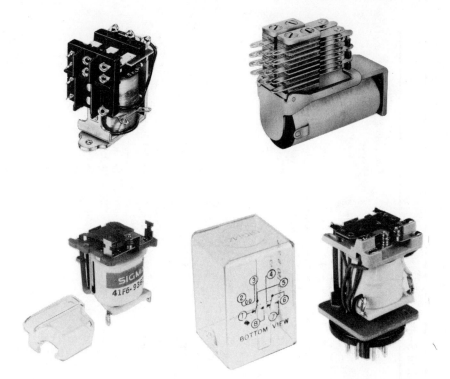

Fig. 7-17 Sensitive electromagnetic relays. Top, Potter & Brumfield general purpose KLT Series and telephone type ML series. Bottom, Sigma Series 41 and 42 relays. (Covers have been removed for clarity.) All relays are shown one-half size. *(Courtesy of Magnecraft Electric Co. and Sigma Instruments Inc.)*

during the shocks. The characteristics of typical sensitive relays are given in Table 7-6.

One-milliwatt sensitivity is about the limit for electromechanical relays. For sensitivity in the microwatt region, hybrid relays must be used. Sensitive hybrid relays combine a solid-state driving amplifier and a reed relay or conventional relay lead switch. See Section 7.15.

The minimum pickup power, or the relay's sensitivity, as published in relay specs, must be taken with a grain of salt. This pickup power is for reference only—a starting point in the design of the relay energizing circuit, particularly for sensitive relays. It is the minimum power that must be applied to the coil to get the relay to operate—slowly, perhaps erratically, but surely. Some amount of overdrive in excess of this power is necessary to get the operate

TABLE 7-6 Typical Sensitive Relays

	Hermetically sealed, open	Hermetically sealed	Open, dust cover, hermetically sealed	Dust cover, or open	Open, dust cover, hermetically sealed	Dust cover
Enclosure Style						
Weight	3 oz., 2 oz.	3.3 oz.	7.3 oz.	4 oz., 3 oz.	3.5 oz.	
Examples	Sigma Series 26 (adjustment CDS)	Sigma Series 22 (adjustment LS)	Sigma Series 5 (adjustment SS)	Sigma Series 48 (adjustment G)	Sigma Series 41 (adjustment S)	Magnecraft Class 62
Contacts	SPDT	SPDT DPDT	SPDT	SPDT	SPDT	SPDT
Contact Rating						
Current (28 vdc)	0.5 A	2 A	0.25 A	2 A	2 A	1 A
Resistance (initial)	75 MΩ	50 MΩ	150 MΩ	50 MΩ	50 MΩ	50 MΩ
Operate Time*	10 to 20 ms	10 to 25 ms	10 to 30 ms	3 to 6 ms	2 to 6 ms	20 ms
Release Time	2.5 ms	5 ms	2.5 ms	–	2 ms	20 ms
Bounce Time	5 ms	2 ms	3 ms	3 to 5 ms	3 to 5 ms	–
Life (operations)						
Mechanical	–		5 million		1 billion	
Rated Load	100.000 ops.	100.000 ops.	10.000 ops.	100.000 ops.	100.000 ops.	–
Coil Power, Rated	–	–				5.4 mw
Pickup Power (min.)	4 mw	12.5 mw	1 mw	10 mw	40 mw	5 mw
Dropout Power (typ.)	1.2 mw	1.5 mw	0.16 mw		3.4 mw	
Operating Temperature	+10 to +40°C	−65 to +125°C	−10 to +40°C	−20 to +70°C	−55 to +70°C	–
Vibration	5 g	10 g	0	5 g	5 g	–
Shock operating	0 g	30 g	0 g	–	–	–
non-operating	100 g	–	100 g	–	100 g	–
Cost: Each, small quantity	–	–	$13–$26	$8–$10	$6–$8	$12–$18

*With overdrive.

time within spec and to get the relay up to meeting shock and vibration requirements. Reed relays must not be overdriven, and there is no point in overdriving a hybrid relay.

7.8 CRYSTAL CAN RELAYS

Crystal can relays (Fig. 7–18) are miniature and subminiature hermetically-sealed high-performance relays originally developed for military and aerospace applications to meet stringent environmental and reliability performance requirements. The relay gets its name from its miniature enclosure configuration, which is identical to the cases of radio-frequency quartz oscillator crystals. In addition to its widespread use in ground-based military and aerospace equipment and space vehicles, aircraft, and guided missiles, crystal-can relays are also used extensively in computers and data-processing systems.

The low moment of inertia of statically and dynamically balanced armatures enables the relay to switch reliably even when subjected to severe vibration and shock. Contacts are gold-plated for long-term performance. Following clean-room assembly, cases are sealed either by soldering or welding. Tungsten inert gas welding is

Fig. 7–18 Crystal can relays, shown full size. Above: Potter and Brumfield SC and SCG series, below: half-crystal can HC and HL series. *(Courtesy of Potter & Brumfield Division of AMF Inc.)*

preferred over soldering because no organic flux is used that may later flake off inside the relay and contaminate the contacts. The construction of a crystal can relay is shown in Fig. 7–19.

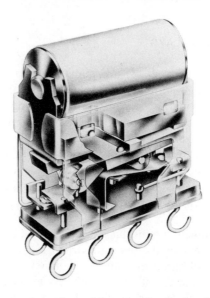

Fig. 7–19 Cutaway view of crystal can relay Series 32 polarized DPDT magnetic latching relay shown. *(Courtesy of Sigma Instruments Inc.)*

The low profile of half-size crystal can relays permit them to be mounted on PC boards having normal 0.5-inch board spacing. Standard 0.2-inch terminal spacing also makes the relay interchangeable with other PC board relays.

If relays are jammed against each other in mounting, operate values can be changed as much as ±25%. To eliminate interaction, space the relays 1/4 inch apart. Operate values will also be affected by stray magnetic fields or nearby permeable iron. Aluminum panel or chassis mounting is preferable to steel.

Crystal can relays are made in several operational configurations: Basic DPDT relay, polarized DPDT, magnetically biased, and polarized DPDT magnetic latching, as shown in Table 7–7.

7.9 DRY REED RELAYS

A dry reed relay, or simply reed relay (Fig. 7–20), is a combination of one or more reed switches (each of which is enclosed in a her-

TABLE 7-7   Typical Crystal Can Relay Characteristics

Contacts: DPDT
Coil Voltages: 6, 12, 24, 48 vdc
Operating Temperature: −65 to +125°C
Storage Temperature: −65 to +150°C
Vibration: 30 g, 10–5000 cps
Shock: 30 to 200 g, varies with relay type and adjustment

	Basic DPDT Relay	Polarized Magnetically Biased	Polarized Magnetically Latched
Case Size	Half	Full	Full, half
Weight	0.4 oz	0.7 oz	0.4, 0.7 oz
Operate Time (Typical)	4 ms	3 to 9 ms	3 to 9 ms
Release Time (Typical)	4 ms	2 ms	–
Transfer Time (Typical)	–	–	0.5 ms
Contact Bounce Time (Typical)	1 ms	1 ms	1 ms
Coil Power, Rated	800 mw	150–450 mw	140–900 mw
Pickup Power (Typical)	260 mw	40–200 mw	20–230 mw
Mechanical Life (Operating)	1 million	Some types to 50 million	2 million to 100 million
Electrical Life (Operations at Rated Load)	100 thousand	100 thousand	100 thousand to 250 thousand
Contact Rating	Dry circuit to 2 A	Low level to 3 A	Dry circuit to 2 A
Capacitance, Contact to Contact	1–2 pF	2–8 pF	2–8 pF
Versions	–	Single coil, Dual coil	Single coil, Dual coil
Cost: Each, small quantity	$10–$20	$21	$21–$35

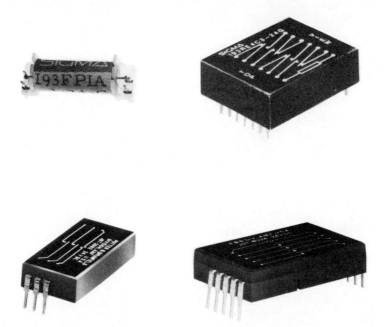

Fig. 7-20 Dry reed relays. Top: Sigma open and enclosed Series 193 reed relays. Enclosed varieties are magnetically shielded. Bottom: Potter & Brumfield JDT 2000 and JDT 4000 two pole and four pole dual-thin line low profile reed relays. Both have magnetic shielding. Relays shown full size. *(Courtesy of Sigma Instruments Inc. and Potter & Brumfield Division of AMF Inc.)*

metically-sealed glass capsule) and an electromagnetic coil. In some versions, biasing permanent magnets are also included. The capsule is either filled with a dry inert gas or evacuated, depending on the end performance requirements. The hollow relay coil surrounds the capsule or group of capsules to make a multipole relay.

Dry reed relays are widely used for counting, selection, and logic in control systems and for isolating input and output equipment interfaces. Dry reed relays are fast-acting and have long life. Being hermetically sealed, they are impervious to environmental contamination and will not suffer the outgassing problems sometimes associated with other types of hermetically-sealed relays because no organic materials are used inside the capsule.

Dry reed relays are made in standard, miniature, and subminiature sizes, as shown in Table 7-8. Enclosure styles include magnetically shielded and encapsulated. Unenclosed dry reed relays are also available. Terminations on most dry reed relays are arranged for PC board mounting; at least one version is encapsulated with an octal type socket. Other dry reed relays have all connections brought out one end in solderless wire-wrap terminals (pages 222–23).

Capsule contact configurations include:

Form A Mechanically Biased
Form B Polarized, Magnetically Biased
Form C Mechanically Biased
Form C Polarized, Magnetically Biased

Form A Mechanically Biased. In this most common SPST–NO reed capsule configuration (Fig. 7–21) two overlapping ferromagnetic reeds (actually slender flat metal blades) are sealed into the ends of a narrow glass tube with their free ends overlapping in the center. Contact areas are either plated, usually with gold or rhodium, or have precious metal contacts inlayed, welded, or brazed.

Since the reeds are ferromagnetic, the extreme ends will assume opposite magnetic polarity when in a magnetic field. With sufficient flux density, attraction between the reeds overcomes their stiffness, and they flex toward each other and make contact. For reliable switching, some overdrive is necessary.

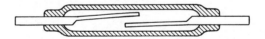

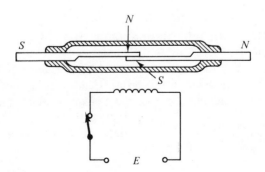

Fig. 7–21 Dry reed capsule, Form A contacts, mechanically biased. *(Courtesy of Magnecraft Electric Co.)*

TABLE 7-8   Typical Data for Single-Capsule Reed Relays†

	Microminiature	Subminiature
Capsule Size (typ.)	0.39 x 0.09 in.	—
Contact Rating		
Form	SPST–NO (Form A)	SPST–NO, DPST–NO, SPDT, SPST–NC, DPDT
Power (max.)	0.1 VA	3 VA
Current (max.)	0.01 A	0.11 A
Voltage, Open Circuit, (max.)	12 vac	28 vdc } for SPST–NO
Resistance (initial)	200 mΩ	200 mΩ
Operate Time (typ.)	0.25 ms	0.2 ms
Release Time (typ.)	0.5 ms	—
Bounce Time (typ.)	0.5 ms	—
Life, Rated Load	10 million operations	10 million operations
Coil Power 1 capsule	300 mw	50–500 mw
(typ.) 2 capsules	—	400–720 mw
3 capsules	—	—
6 capsules	—	—
Coil Voltages, Standard	6, 12, 24 vdc	6, 12, 24, 48 vdc
Enclosures	Tubular, epoxy encapsulated Modular PC package Electrostatic shielding Magnetic shielding	Epoxy, molded DIP
Cost: Each, small quantity	$7	$4 to $6

†All are available (except DIP) in multiple-capsule versions with increased package size.

Form B Polarized, Magnetically Biased. If a Form A switch capsule is biased to the closed position with a small permanent magnet, the result is an SPST–NC contact set (Fig. 7–22). When the coil is energized in the proper polarity, the coil's magnetic field cancels the magnetic field of the biasing magnet, and the contacts open. If excessive voltage is applied to the coil, the magnetic field of the coil will *exceed* the magnetic field of the biasing magnet and will cause the contact to reclose.

Form C Mechanically Biased. This SPDT switch capsule contains one compliant reed and two rigid stationary contacts (Fig. 7–23). Before it is fused into the glass, the movable reed is positioned so that pressure is exerted on the normally-closed contact. To

TABLE 7-8   *Continued*

Miniature	*Standard*	*Standard (Power)*
0.75 x 0.1 in.	2.0 x 0.2 in.	2.0 x 0.2 in.
SPST–NO, DPDT, SPDT	SPST–NO, SPDT	SPST–NO
12 VA	10 VA	100 VA
0.25 A	0.5 A	3 A
100 vac	250 vac	—
150 mΩ	100 mΩ	150 mΩ
1 ms	2 ms	1 ms
0.5 ms	0.5 ms	1 ms
0.5 ms	0.5 ms (NO) 4 ms (NC)	0.5 ms
5 million operations	25 million operations	200 thousand operations (1 A)
600 mw	300 mw	500 mw
900 mw	—	—
—	750 mw	—
—	1500 mw	—
6, 12, 24 vdc	6, 12, 24 vdc	6, 12, 24 vdc
Open style, molded bobbin	Open style, molded bobbin	Open style, molded bobbin
Tubular, epoxy encapsulated	Tubular, epoxy encapsulated	Modular PC package
Modular PC package	Modular PC package	
Electrostatic shielding	Electrostatic shielding	
Magnetic shielding	Magnetic shielding	
$3–$8	$5–$14	$5

limit the magnetic attraction between the normally-closed contacts, the normally-closed stationary contact has a pad of nonmagnetic material welded or brazed to it — performing essentially the same function as a residual screw of a telephone relay.

When the relay coil is energized, the magnetic attraction is stronger between the normally-open contact and the movable contact than it is between the movable contact and the normally-closed contact. This produces the switching action.

Form C Polarized, Magnetically Biased. In this reed switch configuration (Fig. 7–24) the compliant movable reed is similar to that in the mechanically biased switch, but both stationary contacts are identical: There is no nonmagnetic residual pad on the normally-

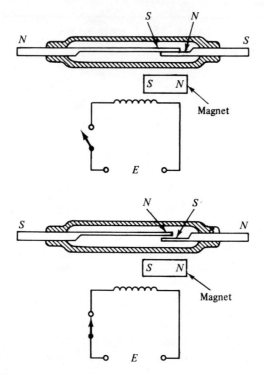

Fig. 7–22 Dry reed capsule, Form B contacts, mechanically biased. *(Courtesy of Magnecraft Electric Co.)*

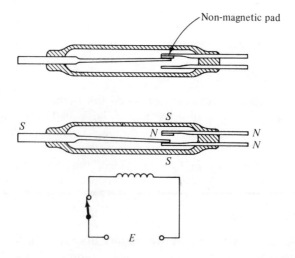

Fig. 7–23 Dry reed capsule, Form C contacts, mechanically biased. *(Courtesy of Magnecraft Electric Co.)*

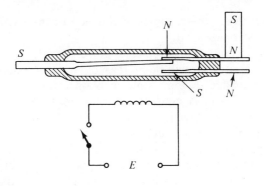

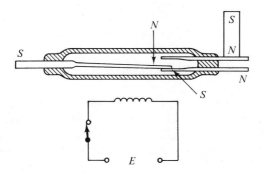

Fig. 7–24 Dry reed capsule, Form C contacts, polarized, mechanically biased. *(Courtesy of Magnecraft Electric Co.)*

closed reed. The movable reed is center–stable; that is, with no magnetic forces applied, it is mid-point between the two stationary contacts. A permanent magnet is positioned to hold the movable reed against one of the stationary contacts.

When the relay coil is energized with the proper polarity, the combination of the compliant contact moves from the normally-closed contact to the normally-open contact.

To minimize contact bounce time in a Form C reed capsule (normally-closed bounce time can double the relay operate time), Form A and Form C capsules can be paired in a relay to provide a Form C SPDT operation.

Operating Considerations. In multiple-capsule dry reed relays, contacts operate in random sequence. Some reed capsules, while within acceptable manufacturing sensitivity tolerances, are simply more sensitive than others and will operate earlier as the flux field builds up. Also, the capsules at the side of the flat pack of capsules are in a stronger field.

Holding circuits using reduced power should be avoided, because contact resistance will be increased. A suppression diode does not significantly affect operate time or bounce time but can increase release time on Form A contacts from 30 μs to 1 ms. Reed relays are easily transistor driven, but a protective reverse-biased diode should be connected across the coil.

Although the reeds are sealed in glass capsules in clean rooms so that foreign particles and active atmospheres are kept out, contact erosion will in time produce tiny magnetic fragments that will naturally collect in the magnetic gap. These fragments can cause intermittent bridging, or what behaves as a "weld" but is not a weld, because there is no molten joining, the contacts being held together by friction alone.

7.10 MERCURY-WETTED REED RELAYS

In this form of reed relay, hermetically-sealed glass capsules similar to those used in dry reed relays are stood on end with a pool of mercury at the bottom. By capillary action, some of the mercury flows up the reed to wet both the movable and fixed reed contact surfaces (Fig. 7–25).

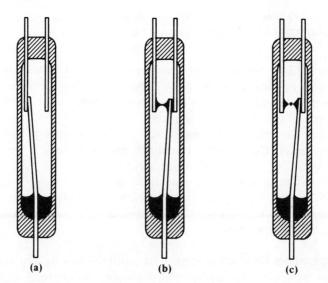

(a) (b) (c)

Fig. 7–25 Mercury-wetted reed contact operation. *(Courtesy of Magnecraft Electric Co.)*

Mercury-to-mercury contact provides several advantages: Consistent operate and release time, absolutely no contact bounce, extremely low contact resistance, and no contact wear. Operation is fast, and the contacts can carry loads dry circuit up to 5 A. The relays have exceptionally long operating life and are capable of high-speed repetitive operation.

The disadvantages are poor resistance to vibration and shock, the necessity to mount and operate the relays in a vertical position, and the fact that contact protective circuits are necessary to achieve long life for all contact loads except dry circuit.

Mercury-wetted relays are made in several styles, as shown in Table 7–9 and Fig. 7–26. Each style enclosure may contain more than one switching capsule. Both the capsules and the surrounding coils are potted. The steel cases provide magnetic shielding and mechanical protection and must not be opened because of the danger of breaking the glass capsule, which is filled with hydrogen under pressure. (See pages 228–29.)

Five contact configurations are used:

Form A Mechanically Biased
Form B Magnetically Biased
Form C Magnetically Biased
Form D (Bridging) Magnetically Biased
Form D (Bridging) Mechanically Biased

Form A Mechanically Biased. The Form A (SPST) mercury-wetted reed capsule has two compliant magnetic reeds. One is sealed into the top end, the other into the bottom end of the capsule.

Form B Magnetically Biased. This mercury-wetted reed switch capsule operates the Form B in the same manner as polarized magnetically-biased dry reed switch.

Form C Magnetically Biased. This mercury-wetted reed switch capsule operates similarly to the Form C dry reed switch. There is no Form C mercury-wetted reed switch in a mechanically biased version.

Form D (Bridging) Magnetically Biased. This SPDT or DPDT make-before-break capsule is similar to the Form C capsule, except that the contact gaps have been spaced closer together so that the mercury film is momentarily drawn into a filament between the movable and closed contact until after the open contact has closed. The filament of mercury then ruptures with the bridging time being 50–500 microseconds.

TABLE 7-9 Typical Mercury-Wetted Reed Relays†

Class	Tubular metal Octal tube plug	Tubular metal Min. tube plug	DIP, epoxy case PC pins	Flat packs PC pins
Weight	4 oz. to 10 oz.	1.5 oz.	—	1.3 oz.
Contacts, Standard	SPDT DPDT 3PDT 4PDT	SPDT	SPST-NO	SPDT DPDT
Contact Forms	Form C Form D	Form C Form D	—	Form C Form D
Contact Rating (max.)	5 A*	2 A*	25 VA* or 1 A*	—
Coil Voltage	—	—	5, 6, 12, 24 vdc	—
Coil Resistance	2 Ω to 25 kΩ	18 Ω to 9200 Ω	—	2.2 Ω to 8600 Ω
Coil Turns	600 to 53,000	900 to 21,000	—	290 to 18,000
Sensitivity (amp. turns)				
Must Operate (typ.)	250	60	—	42
Must Release (typ.)	105	10	—	—
Operate Time (typ.)**	6 ms	—	1.5 ms	—
Release Time (typ.)**	6 ms	—	—	—
Bridging Time (typ)	50 to 500 μs	—	—	—
Frequency of Operation (max.)	60 to 100 ops./sec	200 ops./sec	—	—
Operating Temperature	−38°C to +105°C	−38°C to +105°C	−38°C to +85°C	—
Vibration	10 g (non. op.)	—	—	—
Shock	30 g (non. op.)	—	—	—
Other versions	Dual coil High speed	Dual coil	—	Dual coil
Cost: Each, small quantity	$14–$26	—	$7–$13	—

†Manufacturer's data for flatpacks and DIPs sketchy.   *With required contact protection.   **With overdrive.

Fig. 7–26 Mercury-wetted reed relays. Potter & Brumfield JMA and JMS Series shown two-thirds size. JMS relay must be mounted vertically, not horizontally as shown. *(Courtesy of Potter & Brumfield Division of AMF Inc.)*

Form D (Bridging) Mechanically Biased. This mercury-wetted reed switch capsule is different from the preceding types. Two pairs of contacts project downward from the top of the capsule. The single movable armature reed projects up from the bottom from the pool of mercury. Contact pads are welded to the armature positioned to short together the pairs of normally-closed or normally-open contacts. During the transfer of the armature from one pair of contacts to the other, all contacts are momentarily shorted together, or bridged, by mercury filament. This capsule can switch higher voltages and currents.

Protective Circuit Requirements. For best life, contact protection is required for mercury-wetted relays. However, on dry circuits, such as thermocouple or strain gage circuits, no contact protection is required.

The usual contact protection circuit consists of a resistor and capacitor in series as shown in Fig. 7–27, installed as close as possible to the relay terminals.

7.11 PC BOARD RELAYS

Almost any relay can be a "PC board relay" if the connections are brought out on pin terminals suitably spaced (Fig. 7–28). All such

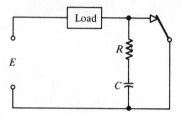

Fig. 7–27 Protective circuit for mercury-wetted relay contacts.

Fig. 7–28 Military and commercial TO-5 PC board relays. Shown one-half size. *(Courtesy of Teledyne Relays)*

PC board relays can be divided into two groups: Those that can be mounted on PC boards stacked on standard 0.6 inch centers, and those with a mounted height greater than 0.4 inch, which cannot. Those relays that can be mounted on standard-stacked PC boards are described in Tables 7–10 (pp. 232–33) and 7–11 (pp. 234–35).

PC board relays include types with conventional coil and armature construction in normal and magnetic-latching versions; in molded, half crystal can, dust cover, TO–5 and ultra-miniature hermetically-sealed cases; and dry reeds in molded, open, and DIPs. Mercury-wetted reed relays are available. There is also a variety of hybrid and solid-state relays in dual-thin-line and DIPs. Dry reeds are available with optional magnetic or electrostatic shielding or with transient suppression diodes. Mercury-wetted reed relays must, of course, be mounted vertically.

TO–5 case relays have all-welded construction. Contact areas are larger than those of conventional miniature relays, and high force-to-mass ratios provide resistance to shock and vibration. Some versions contain discrete diodes for transient suppression. Besides normal actuations, they are available in sensitive and magnetic-latching versions.

Ultra-miniature types are essentially the same as TO-5 types except that leads are arranged on a PC grid rather than in a circle. This arrangement eliminates the need for lead bending and spreader pads for mounting.

7.12 STEPPING RELAY

In a stepping relay, also called an impulse relay or sequence relay, the operation of a single group of contacts is programmed by cams actuated by the relay armature (Fig. 7–29).

According to NARM definition, an electromagnetically-operated stepping switch is not considered to be a form of a stepping relay. However, not everyone has gone along with this definition, including some relay manufacturers who also make rotary stepping switches.

The stepping switch has multipole and multiposition contacts arranged in a semicircular bank. As the armature is driven, contact positions are sequentially closed, opened, or connected or disconnected to the armature contact a step at a time. Stepping switches are covered in Chapter 10.

On the other hand, the stepping relay has one bank or level of contacts arranged in one or more pileups. Each pileup is actuated by a separate cam, but all the cams rotate together. The cam shaft is ratcheted by the relay armature. On special order, a limited amount of special programming can be built into the cams.

Stepping relays are used for applications requiring on–off or reversing action on alternate operations. Relays are rated for intermittent duty or pulse operation only. Contacts, from SPST–NO to

TABLE 7-10 PC Board Relays—Typical Data

	Relay Type		
	Dry Reed	Mercury-Wetted Reed	Conventional
Enclosure Style	DIP	DIP	Molded
Weight	0.1 oz	0.3 oz	0.3 oz
Contacts Available	SPST–NO, NC, DPST–NO, SPDT, DPDT	SPST–NO	SPST–NO, SPDT, DPDT
Contact Rating (dc)			
Power (max.)	3 to 6 VA	25 VA	–
Current (max.)	110 to 500 mA	1 A	2 A
Resistance (initial)	150 to 500 mA	100 mΩ	50 mΩ
Operate Time	0.2 to 1 ms	0.5 ms	3 ms
Release Time	0.1 ms	–	2 ms
Bounce Time	Included	0	2–6 ms
Life			
Dry Circuit	100 million ops.	–	10 million ops.
Rated Load	1 to 10 million ops.	40 million ops.	1 million ops.
Coil Power, Rated*	50 to 160 mw	500 mw	125 to 450 mw
Pickup Power (max.)*	30 to 108 mw	260 mw	–
Coil Voltage	5, 6, 12, 24 vdc	5, 6, 12, 24 vdc	5, 6, 12, 24, 48, 110 vdc
Operating Temperature	−55°C to +85°C	−38°C to +85°C	−55°C to +65°C
Vibration	20 to 100 g	Not rated	–
Shock	50 to 100 g	Not rated	20 g
Cost: Each, small quantity	$4–$7	$9–$10	$4

*For single pole relays.
**SPST 8 A also available.

4PDT, are silver buttons rated 5 A at 24 vdc or 115 vac. Hermetically-sealed and dust cover enclosures are available. Table 7–12 details typical stepping relays (page 236).

7.13 COAXIAL RELAYS

Coaxial relays (Fig. 7–30) are used for radio-frequency switching operations between coaxial-cable connected equipment. They are

TABLE 7-10 *Continued*

Relay Type			
Crystal Can	*Dry Reed*	*Conventional*	*Conventional*
Metal, hermetically sealed	Molded or open	Dust cover	TO–5
0.4 oz	0.14 to .7 oz	0.5 to 1.1 oz	0.09 oz
DPDT	SPST–NO, DPDT, 3PDT, 4PDT, 8PST–NO	SPDT, DPDT, 4PDT, 6PDT	DPDT
—	3 to 10 w	—	—
2 A	—	1 A**	1 A
50 mΩ	100–200 mΩ		100 mΩ
4 ms	2 ms	8 ms	4 ms
4 ms	2 ms	6 ms	3 ms
1 ms	2 ms	—	—
1 million ops.	—	100 million ops.	10 million ops.
100 thousand ops.	—	6 million ops.	100 thousand ops.
1.2 w	350 mw	450 mw	450 mw
260 mw	220 mw	250 mw	185 mw
6, 12, 24, 48 vdc	3, 6, 12, 24 vdc	6, 12, 24, 48 vdc	5, 6, 9, 12, 18, 26.5 vdc
−65°C to +125°C	—	to 70°C	−55°C to +71°C
30 g	—	10 g	10 g
50 g	—	12 g	30 g
$12	$3–$12	$4–$9	$11

used extensively for antenna switching in mobile, aircraft, and marine two-way radios. These relays have switching contacts enclosed in a metal cavity that is dimensioned to provide a close match to the characteristic impedance of the interconnecting cable used and electrical shielding. This switch configuration introduces minimum loss.

Coaxial relays feature low voltage standing-wave ratio (vswr), excellent cross talk characteristics, long life (one million cycles, minimum), fast operation, high reliability in small size at low cost, and a wide selection of connectors. Some have optional auxiliary contacts, mechanical latching, and sensitive coils.

TABLE 7-11   PC Board Hybrid and Solid State Relays—Typical Data

	Relay Type		
	Hybrid		Light-Emitting Diode, Opto Isolator and Zero-Crossover Turn on Triac
	Reed Triggered Triac	Amplifier Driven Reed	
Example	Potter & Brumfield JDB Series	Potter & Brumfield JDA Series	Potter & Brumfield JDO Series
Enclosure	Dual thin-line	Dual thin-line	Low-profile
Weight	0.5 oz	0.4 to 0.7 oz	0.6 oz
Contacts	SPST–NO	DPST–NO	SPST–NO
Switching Device	Triac	Reed Relay	Triac
Switching Device Load Rating	1.7 A 120 vac	0.5 A low	1.7 A 120 vac
Surge (one cycle)	20 A	–	20 A
OFF Leakage Current	1 mA	–	1 mA
Operate Time	1 ms	3.5 ms	–
Release Time	10 ms	10 ms	–
Life (operations)	10 million	1 to 10 million	10 million
Control (trigger)	300 mw	2.4 v 100 μw	24 vdc 170 mw
Power Supply (typical)	None req'd.	24 vdc 500 mw	None req'd.
Cost: Each, small quantity	$13	$17	$17

Fig. 7–29 Stepping relays. Above: Automatic Electric type OCS, below, Potter & Brumfield GM and AP series. *(Courtesy of GTE Automatic Electric and Potter & Brumfield Division of AMF Inc.)*

TABLE 7-11   *Continued*

Relay Type

Solid State

Transformer-Coupled Semiconductor

Teledyne	Teledyne	Teledyne
640-1	641-1	644-1
DIP	DIP	DIP
0.2 oz	0.2 oz	0.2 oz
SPST–NO	SPST–NO	SPST–NO
Bipolar	Triac	Transistor
80 mA	1 A	400 mA
50 vac/dc	140 vac	60 vdc
0.1 joule	10 A	–
60 μA	1 mA	60 μA
1 μs	20 μs	1 μs
10 μs	8.3 ms	5 μs
No known wearout mode	No known wearout mode	No known wearout mode
10 vdc	10 vdc	10 vdc
–	15 mA	15 mA
$11	$13	$14

Coaxial relays are built in two basic styles and in several sizes. One style is with RF connectors as an integral part of the cavity; the other is with shielded cables connected directly to the cavity.

Coaxial relays consist of a cavity and an actuator. The cavity housing is a die casting with a cover plate that is removable for inspection of the contacts. Cavities are made in several physical layouts to accommodate a variety of external lead layouts, because coaxial cable is not that flexible. Contacts are gold-plated silver cadmium oxide and are supported directly from the cable or connector center conductors.

Actuators are, roughly speaking, telephone-type relays minus the contact pileups or with one pileup serving as auxiliary contacts.

The integral-connector type coaxial relays are suited for heavy-duty panel mounting and high-power applications. These versions

TABLE 7-12   Typical Stepping Relays

	Automatic Electric Series OCS	Potter & Brumfield AP Series	Struthers–Dunn 211 Frame
Steps per Revolution	30, 32, 36	4	6, 8
Number of Cams	1 to 8	2	2
Switching Contacts per Cam (maximum)	6 springs	SPST–NO to 4PDT	DPDT
Operation	Indirect drive	Indirect drive	Indirect drive
Contact Rating	3 A (135 w)	5 A 120 vac	5 A 115 vac
			5 A 24 vdc
			5 A
Speed:			
Self-Interrupted	65 steps/sec	Not given	Not given
Impulse-Controlled	30 steps/sec		
Coil Voltage	24, 110 vdc	6, 12, 24, 48, 110 vdc	6, 12, 24, 32, 115 vdc
	110 vac (with rectifier)	6, 12, 24, 120 vac	12, 24, 115, 230 vac
Coil Power	—	9 w	6 w
		19 vA	12 vA
Cost: Each, small quantity	$23 (3 cams)	$17	—

*Blank cams available.

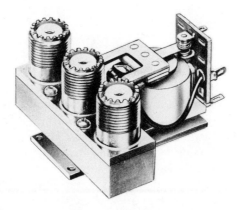

Fig. 7–30 Coaxial relay with integral connector type cavity and auxiliary contacts. Magnecraft Class 128 relay shown can switch 100 watts up to 4000 MHz. Relay is shown half-size. *(Courtesy of Magnecraft Electric Co.)*

terminate in type UHF, N, BNC, or TNC connectors. Typical coaxial relay characteristics are shown in Table 7–13.

TABLE 7-13   Typical Coaxial Relays (Magnecraft)

RF Contacts			
Load Rating	100 w	150 w	1000 w
Contact Form	SPDT	SPDT	SPDT
Maximum Frequency	4000 MHz	470 MHz	470 MHz
VSWR (0 to 470 MHz)	1.25:1	1.25:1	—
Crosstalk (0 to 470 MHz)	−40 db	−40 db	—
Auxiliary Contacts			
Contact Form	SPDT, DPDT	SPDT	None
Current Rating	10 A	5 A	None
RF Connection Std.	UHF	RG–58 A/U Cables 12 in. long	UHF —
Available	N, BNC, TNC		
Operate Time	20 ms	15 ms	25 ms
Release Time	20 ms	7 ms	10 ms
Coil Voltages	6, 12, 24, 48 vdc 120 vac	6, 12, 24, 48 vdc	6, 12, 24 vdc 115 vac
Coil Power, Rated	1.5 w	1.0 to 2.5 w	2 w
Operating Temperature	−55°C to +65°C	−55°C to +65°C	−55°C to +85°C
Cost: Each, small quantity	$23	$11	$35

Coaxial Relay Terms.

VOLTAGE STANDING-WAVE RATIO (vswr): The power loss, which is due to mismatch introduced into the line by a relay or any other device, expressed as a ratio of the highest voltage to the lowest voltage found in the RF line. In RF transmission lines, it is desirable to avoid, wherever possible, standing waves that create reflected power losses.

INSERTION LOSS: The difference between the power received and that transmitted, before and after the insertion of, in this case, *closed* relay contacts into the transmission line. The attenuation is measured in dB.

CROSS TALK: Interference caused by stray electromagnetic or electrostatic coupling of energy between circuits — in coaxial relays, the coupling between a closed circuit, or other open or closed contacts on the same relay switch. It is expressed in dB from the reference signal level.

GROUNDING: Open contacts in a coaxial relay may be either grounded or terminated. In a grounded coaxial relay, open contacts are grounded against the cavity wall. Grounding contacts reduce RF leakage to the disengaged circuit and are recommended where optimum circuit isolation is required. Cross talk with grounding can be 20 dB lower than in terminated circuits. In nongrounding coaxial relays, the disengaged contact blade is terminated with a resistor.

7.14 METER RELAYS

Meter relays, also called instrument relays, use modified D'Arsonval meter movements as the actuator for the contacts. There are three basic types of meter relays: *Locking coil,* in which there is physical contact between the moving meter pointer and the switching contacts; *magnetic,* in which the meter pointer also physically closes contacts but is held by a magnet and must be mechanically or electromagnetically reset; and *optical,* in which there is no physical contact between the moving pointer assembly and the controlled contacts.

Instrument relays are extremely sensitive — in the order of microwatts. They have pickup precisely adjustable over a wide range, and they allow visual monitoring of the circuit. Meter relays are used extensively in industrial control and monitoring. Table 7–14 compares the characteristics of the three types of meter relays.

Locking-Coil Meter Relays. This type of relay (Fig. 7–31) is used to indicate, monitor, or control variables that can be measured and translated into electrical units including pressure, temperature, pH, revolutions per minute, roentgens, mechanical position, torque, timing, sound, light, etc. Locking-coil meter relays are also used for limit-control applications and can be wired to sound an alarm or to stop machinery or a process.

The meter relay uses a modified taut-band or jewelled D'Arsonval meter movement as an actuator. The conventional signal coil with attached indicating pointer rotates in the flux gap of a permanent magnet. When the coil is energized, a torque is produced, causing the pointer to move proportionately.

The meter relay also has an adjustable set–point arm, two contacts, and a locking coil. The set–point arm can be manually posi-

TABLE 7-14   Meter Relay Types

	Locking Coil	Magnetic Contact	Optical
Ranges	0–5 μA to 0–9 Adc 0–3 mv to 0–500 vdc ac ranges available	2–0–2 Adc 1–0–1 mAdc 0–5 μA to 0–1 mAdc 0–1 mAac, 0–97 mAac 0–10 mvdc, 0–50 mvdc	0–5 A to 0–30 Adc 0–10 A to 0–50 Aac 0–5 mv to 0–500 vdc 0–250 mv to 0–500 vac
Lamp Excitation	–	–	2 vdc, 270 mA 6, 12, 24 vdc lamps special order
Reset Voltage	–	6 vdc std., 12, 24, 48, 120 vdc, 115 vac available	–
Accuracy, Full Scale	±2% dc	±5% dc	±2% dc
Deflection	±3% ac	±5% ac	±3% ac
Repeatability	1% fsd	1% fsd	–
Coincidence Error	–	–	$1\frac{1}{2}$%
Set Point Resolution	–	–	$\frac{1}{2}$%
Operating Temperature	Room ambient	Room ambient	Room ambient
Contact rating	10 to 25 mA	100 mA	10 mA
Cost: Each	$63–$171	$59–$140	$61–$125

Fig. 7–31   Locking-coil meter relay. *(Courtesy of LFE Corporation)*

tioned by a knob on the front of the meter so that relay actuation can occur at any desired meter current. The set–pointer carries a spring contact connected to one side of the meter input. The other contact is mounted on the indicating pointer in series with the locking coil. When the signal reaches set–point, the two contacts "make."

The locking coil is wound on the same coil form as the signal coil. It is part of the moving element and is connected to the indicating pointer contact. The making of the meter contacts energizes the locking coil. This increases the torque developed by the signal coil and drives the contacts together forcefully. The locking-coil current holds them firmly closed to create a low-resistance circuit.

Making of the contacts loads the flexure-spring contact on the set–point arm. When the locking-coil circuit is broken (manually or automatically), the spring contact unloads, kicking the contacts apart without teasing or sticking. As opening the circuit takes place after the locking-coil circuit is broken, contacts are not subject to arcing.

The meter relays will take current overloads up to 10 times the full scale current without *mechanical* damage to the meter. Electrical damage is something else, because a 10 times overload can quickly burn out hairsprings, coil, windings, series resistors, or shunts. A stabistor diode should be used to prevent damage.

The two coils in a standard locking-coil meter relay are electrically connected. These meters can be provided with isolated coils which are generally required in any application in which other instruments operate from the same signal source as the meter relay. This prevents the locking-coil current from upsetting the accuracy of external instruments in the signal circuit.

Meter relays can also be provided with a double locking coil for double set–point meter relay. The two locking coils are connected so that one locks up-scale and the other locks down-scale. The coils are wound in opposite directions to lock the contacts, at the high and low set–point, with the same polarity.

The response time of meter relays is approximately the same as that of standard indicating meters, but response times from 300 ms to 10 s can be obtained.

Magnetic-Contact Meter Relays. These relays (Fig. 7–32) provide a simple but positive control action. When the moving element contacts close at the set–point, auxiliary contact pressure is provided by small permanent magnets to insure a low-resistance contact circuit. Unlocking must be accomplished by an electromechanical or mechanical reset mechanism. A magnetic-contact meter relay will switch either dc or ac.

Fig. 7–32 Magnetic-contact meter relay. *(Courtesy of LFE Corporation)*

Set–points of magnetic contact meter relays are usually factory-set, but some models have adjustable set–points. Most magnetic-contact meter relays can be obtained with single or double set–points.

Optical Meter Relays. Optical meter relays (Fig. 7–33) provide noncontacting control. These meter relays may be used to control any variable that can be converted to analog voltage or current signals from ac or dc devices, such as thermocouples, resistance temperature detectors, or other transducers.

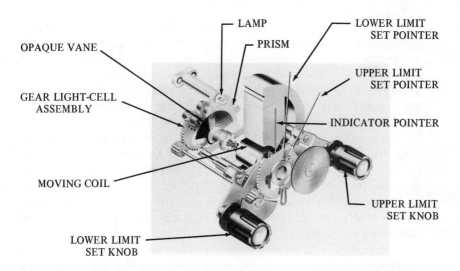

Fig. 7–33 Optical meter relay. *(Courtesy of LFE Corporation)*

The optical meter relays provide one or two visible set–points and meter indication of the control variable. These relays use a modified D'Arsonval meter movement. A D'Arsonval meter shutter or vane is connected to and rotates with the moving coil and indicating pointer. A photoconductive cell is connected to the set–point indicator. As the signal coil and indicating pointer rotate, the vane or shutter moves between the cell and a light source, causing a change in the amount of the cell and providing the set–point signal for alarm or control circuits.

7.15 HYBRID AND SOLID-STATE RELAYS

These relays (Fig. 7–34) combine diverse combinations of load-switching and triggering devices in order to obtain (1) sensitivity, (2) isolation between control circuit and load, or (3) economical high power-handling capability. There are two types of hybrid relays: Solid-state switching devices driven by isolating relays, optical couplers, transformers (Fig. 7–35), etc.; and conventional and reed relays driven by solid-state amplifiers. Solid-state relays consist of a solid-state switching device driven by either an amplifier, an LED optical coupler, or a transformer.

The magnitude of the load current, and whether it is ac or dc, determines the choice of semiconductor switching device. For low-level dc switching, FETs are used; bipolar devices are used for intermediate and power level dc switching; triacs are used for ac or dc power switching.

The *triac* is a three-terminal semiconductor with ON–OFF switching characteristics similar to that of a silicon controlled rectifier except that it provides switching regardless of the polarity of the applied voltage. The triac is controlled by a single gate, and in ac applications, *zero crossing* switching can be performed. (In zero crossing switching, the triac is switched at the instant the ac load voltage goes through zero, thus minimizing EMI and RFI.)

If used properly, solid-state relays have high reliability with long life and no known wear-out modes. Operation is silent, and there is no contact bounce or arcing, making operation relatively transient-free. They are better at handling reactive loads and are unaffected (within reason) by shock and vibration. Since they have no physically opening and closing contacts, they are adaptable to hazardous applications.

Replacing a conventional relay with a solid-state relay is not a simple one-for-one trade, however. It is unlikely that you will end up with a smaller package size, even when excluding the heat sink the

Fig. 7–34   Solid-state and hybrid relays. **(a)** Series EBA amplifier-driven thyristor, **(b)** Series 640-1 transformer-isolated bipolar transistor, **(c)** Series 611 dc controlled optically-isolated thyristor ac load relay, **(d)** Series KUA amplifier-driven conventional relay, **(e)** Series EBT reed relay-driven thyristor, **(f)** Series JDA amplifier-driven reed relays. All of the relays are shown half-size. *(Courtesy of Potter & Brumfield Division of AMF Inc. and Teledyne Relays.)*

solid-state relay will need. Excitation must be clean and must be free of surges and transients. Sudden line fluctuations can turn on the relay even without a gate signal. Solid-state relays are single-pole devices. ON resistance is measured in ohms, not milliohms, and significant leakage current flows in the load circuit when it is switched OFF. Exceeding load current ratings in conventional relays usually results only in shortened life through accelerated contact erosion; overloading a solid-state relay just once can result in catastrophic failure. There are ambient temperature limitations. There is also the matter of cost: With a conventional relay, you can switch 10 A for

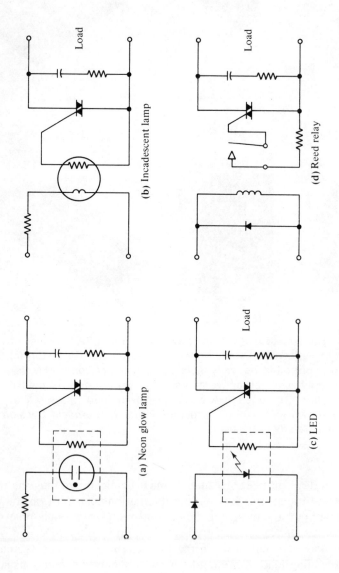

(a) Neon glow lamp

(b) Incadescent lamp

(c) LED

(d) Reed relay

Fig. 7-35 Triac relay coupling circuits.

$3 – go solid-state, and the cost jumps to $17–$28. Hybrid relays are varied in design and application. Characteristics are compared in Tables 7–15 and 7–16.

Triac Hybrid Relays.   These relays inexpensively provide isolation between a low-power control circuit and a high-power load. The devices use either an opto-isolator, a reed or conventional relay, or a solid-state amplifier to sense and provide the gating to a triac. The opto-isolator signal source can be either an incandescent lamp or a light-emitting diode. The units are packaged in rectangular nylon cases, or in flat packs for PC board mounting, usually color-coded to indicate load switching capability.

Monitors.   Sensitive monitoring relays combine silicon solid-state amplifiers with miniature dry reed load contacts. Versions can provide overvoltage and undervoltage sensing, with either automatic reset or electrical latching.

Modules contain a solid-state amplifier driving a dry reed or conventional load switching relay. A dc supply and a dc input signal voltage are required.

Typical applications for these relays include: Alarm circuits, fault detection, frequency selection, memory logic, metering, multiple-gate logic, and pulse stretching.

Micro-Sensitive Relays.   These solid-state relays operate from signals in the microwatt range, a level far below the most sensitive electromechanical relays.

AC Null-Sensing Relays.   These sensitive ac-operated devices contain two sets of reed switch load contacts that respond to phase reversals of the input signal. These hybrid modules can replace more complicated and more expensive components commonly used for sensing, regulating, and positioning in-process and machine control systems.

Both reed switches are open with the signal system at null. An unbalance beyond the null range in either direction will close one or the other switch, depending upon the direction of unbalance.

7.16 WHO MAKES WHAT

Table 7–17 lists some of the major relay manufacturers.

TABLE 7-15 Typical Triac Hybrid Relays

Load Current:	1.7 A to 20 A, dependent on triac used
Load Voltage:	120, 240 vac
"Switch Contact":	SPST–NO
OFF Leakage Current:	5 ma

Surge Current:	100 A (one cycle)
Operate Time:	2 ms
Release Time:	9 ms

	Triggering Device				
	Incandescent Lamp & Opto Isolator	Conventional Relay	Reed Relay	Light-Emitting Diode & Opto Isolator	Amplifier Driven
Class	Hybrid	Hybrid	Hybrid	Solid state	Solid state
Examples	Sigma 223	Sigma 221	Potter & Brumfield EBT	Sigma 223	Potter & Brumfield EBA
Enclosure	Molded	Molded	Molded	Molded	Molded, or dust cover
Mounting	Flange	Flange	Flange, PC board	Flange	Octal socket, stud
Terminals	Quick connect tabs	Quick connect tabs	PC pins	Quick connect tabs	Quick connect tabs, octal plug
Weight	4.5 oz.	4.5 oz.	4.5, 6.5 oz.	4.5 oz.	6 oz.
Triggering Power	240 mw	360 mw ac or dc	300 mw	50 mw dc only	60 µw
Power Supply (Continuous)	None	None	None	None	24 vdc 1.2 w
Cost: Each	$10	$11	$17–$22	$10	$25

TABLE 7-16   Typical Amplifier-Driven Hybrid Relays

	Solid State Driver	Integrated Circuit Driver	Sensitive Monitor Relay	Null Sensing	Micro Sensitive Monitor Relay
Examples	Potter & Brumfield KUA series	Potter & Brumfield JDA series	Struthers–Dunn Models 11, 12	Struthers–Dunn Models 14, 15	Struthers–Dunn Model S 16
Enclosure	Plastic dust cover	Dual thin line	Epoxy encapsulated	Epoxy encapsulated	Epoxy encapsulated
Mounting	Stud	PC board	PC board & stud	Octal socket	PC board
Terminals	Quick-connect, solder, PC	PC	PC	Octal plug	PC
Weight	3.5 oz.	0.4 to 0.7 oz.	0.8 oz.	—	0.3 oz.
Switching Device	Conventional relay	Reed relay	Reed relay	Reed relay	Reed relay
Contacts	SPST-NO DPDT	DPST-NO	SPST-NO	DPST-NO	SPST-NO DPST-NO
Contact Rating	5 A, 10A	10 w 0.5 A	0.5 A 250 v	0.5 A 250 v	0.5 A 135 v
Operate Time, max.		—	10 ms	10 ms	1 ms
Release Time, max.		—	5 ms	—	0.5 ms
Life Expectancy, Operations at Rated Load	100 thousand	Not given	10 million	10 million	10 million
Triggering Signal	60 µw 2.4 vdc	100 µw 2.4 vdc	1 mw 1 vdc	— 0.5 vac	5 A —
Power Supply					
Voltage	24 vdc	24 vdc	8 vac, 50–400 Hz	120 vac, 50–400 Hz	24 vdc
Power	—	500 mW	1 ma	—	—
Power, Relay Energized	1.2 w	20	20 ma	—	8 mw
Cost: Each, small qty.	$15	$17	$24–$27	$32, $47	$10–$15

TABLE 7-17 Relay Manufacturers

	Allied Controls	Amphenol	Arco-Siemens	LFE Corp.	Automatic Electric	Guardian	Magnecraft	Ohmite	Potter & Brumfield	Sigma	Struthers-Dunn	Teledyne
General-Purpose												
Standard	X		X			X	X	X	X	X	X	
Small			X			X	X		X	X	X	
Miniature			X						X	X		
Power, All Kinds						X	X		X	X	X	
Telephone												
Standard					X							
Medium							X		X			
Small					X		X		X			
Miniature					X		X		X			
Subminiature							X					
Card-Actuated												
Industrial			X		X		X		X	X	X	
Wire Spring					X							
Sensitive												
Hermetically Sealed										X		
Dust Cover							X		X	X	X	
PC Flat Pack										X		
Crystal Can, All Kinds	X								X	X	X	
Dry Reed												
Standard					X		X		X		X	
Miniature					X		X		X		X	
Subminiature							X		X		X	
Microminiature							X					
DIP							X		X		X	
Mercury-Wetted Reed												
Octal									X		X	
Miniature						X			X			
PC Flat Pack									X		X	
DIP							X					
PC Board												
Conventional			X				X		X	X		X
Dry Reed							X			X	X	
Mercury-Wetted Reed							X					
Hybrid									X			X
Stepping, All Kinds					X	X			X		X	
Coaxial, All Kinds	X						X					
Meter (Instrument)												
Locking-Coil				X								
Magnetic Contact				X								
Optical				X								

TABLE 7-17   *Continued*

	Allied Controls	Amphenol	Arco-Siemens	LFE Corp.	Automatic Electric	Guardian	Magnecraft	Ohmite	Potter & Brumfield	Sigma	Struthers-Dunn	Teledyne
Triac Hybrid												
Conventional Relay										X		
Reed Relay								X	X			
Opto-Isolator									X	X		
Amplifier									X			X
Zero-Crossover									X			
Amplifier-Driven Hybrid, All Kinds									X			X

SUGGESTED READINGS

Bursky, Dave, "Focus on Miniature Relays." *Electronic Design 11,* May 24, 1975.

Definitions of Relay Terms. National Association of Relay Manufacturers, Scottsdale, Arizona, February 1975.

Designers' Handbook and Catalog of Reed and Mercury-Wetted Contact Relays, 2nd ed. Magnecraft Electric Co., Chicago, Illinois, April 1967.

Dilatush, Earle, "Relays with Low Profiles and PC Terminals, They Do Many New Jobs." *EDN,* April 5, 1975.

Gilder, Jules H., "Solid-State Relays are Finding Gradual Acceptance—At Last." *Electronic Design 19,* September 13, 1973.

Grossman, Morris, "Focus on Reed Relays." *Electronic Design 14,* July 6, 1972.

Johnson, Phillip, "Solid-State Relays—A Guide to Their Design and Application." *EDN,* October 5, 1973.

National Association of Relay Manufacturers, *Engineers Relay Handbook.* Hayden Book Co., Rochelle Park, N. J., 1966.

Selection Guide to AE Relays and Switches. GTE Automatic Electric, Northlake, Ill.

CATALOGS

LFE Corp., Waltham, Mass.
GTE Automatic Electric, Northlake, Ill.
Bunker Ramo Amphenol, Broadview, Ill.
Magnecraft Electric Co., Chicago, Ill.
Potter & Brumfield Division, AMF Inc., Princeton, Ind.
Sigma Instruments, Inc., Braintree, Mass.
Struthers–Dunn, Inc., Pitman, N. J.

CHAPTER EIGHT

Time Delay Relays

Time delay relays (Fig. 8–1) provide a desired time interval between an actuating signal and the operation of a load-switching device. The actuating signal may be the application of voltage, which can be either steady-state or a pulse; or it may also be the removal of a voltage, even momentarily. For some types, operating power is required in addition to the signal voltage. The load-switching device is either an integral electromagnetic relay of conventional design, a reed relay, or one or more snap-action switches.

Time delay relays are called many things, depending on their function. The term *time delay relay* is used to generically identify the component and also to identify one of the four functional classes (Table 8–1). Figure 8–2 shows the operating functions graphically.

Time Delay Relay. A time delay relay in which the load circuit is energized a desired time interval after the relay control circuit is energized. (Delayed ON.)

Delay Timer. A time delay relay in which the load circuit is energized when the relay control circuit is energized. It remains energized for a specific period of time after the relay control circuit is de-energized.

Interval Timer. A time delay relay in which the load circuit is energized when the relay control circuit is energized. It remains energized for a specified period of time.

251

Fig. 8–1 Motor-driven and solid-state time delay relays. **(a)** Cramer 402/403 reset timer can be used for interval or delay timing, comes in thirteen timing ranges from 6 seconds to 24 hours. **(b)** Cramer model 3810 industrial time delay relay comes in five standard time ranges to 5 minutes, has solid-state timing circuitry, conventional relay for load switching. Both are shown half-size. *(Courtesy of Cramer Division, Conrac Corp.)*

TABLE 8-1   Classes of Time Delay Relays and Some Commonly Used Alternate Nomenclature

Time Delay Relay	Delay Timer	Interval Timer	Repeat Cycle Timer
Delay on Energization	Delay on de-energization	Automatic reset timer	Cycle timer
Delay on Operate	Delay on dropout (DODO)	Interval/momentary	Flasher
Delay on Pull-In (DOPI)	Delay or release timer	Interval/momentary start	Free running timer
Delay on Sustained Contact	Off-delay timer	Momentary start timer	Intervalometer
Delay Timer	Pulsed timer	Period timer	ON–OFF Timer
On–Delay Timer	Reset timer	Pushbutton timer	Recycling timer
Reset Timer	Timed latch		Sequence timer
Time Delay on Operate			
Time Delay Timer			

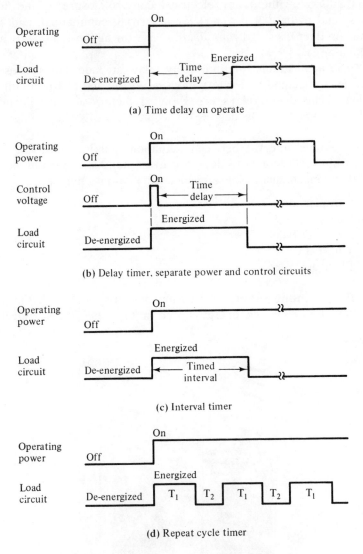

Fig. 8-2 Time delay relay operating modes.

Repeat Cycle Timer. A time delay relay that repeats an ON–OFF timing cycle as long as the relay control circuit is energized.

Time Delay Relay Classes. Calling a time delay relay a *timer* leads to confusion, because the term *timer* is ambiguous. A timer is also a device that provides an audible or visual signal at the end of the timed interval without any electrical contact closure. (Time delay relays tend to be called *timers* for industrial, commercial, and consumer product uses and *time delay relays* for military and aerospace applications.)

The applications for time delay relays include data processing operations, machine tool safety device control, and alarm circuit actuating. Time delay relays are supplied factory-set or adjustable. Each type varies in its degree of accuracy, variety available, and cost.

The time interval between actuation and load-switching can range from milliseconds to days and even months, depending on the time delay mechanism. These mechanisms fall into five classes:

Escapement
Dashpot
Motor-Driven
Thermal
Electronic

Escapement Timer. An escapement, or clockwork time delay relay (Fig. 8–3), has energy stored in its spring. When triggered,

Fig. 8–3 Escapement time delay relay. Setting knob winds spring. Device is shown half-size. *(Courtesy of M. H. Rhodes, Inc.)*

the escapement-controlled gear train operates a cam which provides mechanical actuation of a switch after a predetermined interval. Winding can be manual or electrical. Escapement timers are not widely used in industrial applications; however, they do have an advantage of not requiring electrical energization while timing and their timing cycle is not affected by power interruptions.

Dashpot Timer. A dashpot timer is a time delay relay that uses either air or liquid which flows under pressure through a small orifice to retard the movement of a moving part of a switch. This produces a time interval before actuation.

The mechanism usually consists of a cylinder with a close-fitting piston and an orifice that allows the air or liquid to leak into or out of the cylinder at a controlled rate. The relatively slow transfer of the air or fluid volume retards the movement of the piston. At the end of its travel, the piston actuates a load switch, thus producing a time delay when the relay is energized or de-energized. On the return stroke, a check value allows the cylinder to fill or empty quickly, thus resetting the relay in a minimum of time. The dashpot may be actuated by a solenoid or spring pressure.

Motor-Driven Time Delay Relay. The motor-driven or electromechanical timer uses a synchronous ac motor or constant-speed dc motor to drive a clutch, gear train, and cam, which operates a switch after a predetermined period.

The electromechanical timers are the most widely used industrial timing devices today; they are also used in military and aerospace applications. Costs are generally low. *One* timer can provide complex programs of time intervals for automatic equipment start-up cycle control or for sequential control of operations, such as those required for a washing machine. Electromechanical timers can provide timing intervals measured in days and months; although they can time down to 0.1 second, they are seldom used below one second.

Thermal Time Delay Relay. Thermal time delay relays (Fig. 8-4) utilize a heating element and a bimetallic or thermally expanding member that actuates linkage and contacts, which are usually snap-acting. When energized, the sensing element is heated and expands or bends, performing the desired switch action. Timing is a function of the rate of heat transfer.

Fixed and adjustable time delays can be obtained over a wide range. Accuracy depends primarily on the sophistication of the design. Temperature compensation can maintain timing accuracies over

BIMETALLIC
MEMBER

CONTACTS

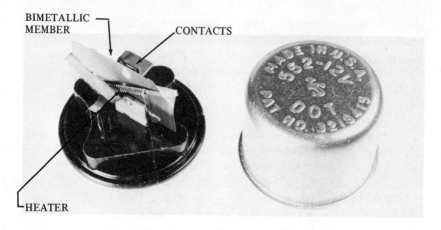

HEATER

Fig. 8–4 Thermal time delay relay. Type shown is used as a repeat cycle timer to control automobile turn signals. Device is shown full size.

a wide range of ambient temperatures. They are not used much any more in precision applications, having been displaced by electronic time delay relays, but they are used by the millions in such applications as automobile turn signals. They are cheap, rugged, sealed and tamper-proof.

Electronic Time Delay Relay. The transistorized electronic time delay relay was originally designed for aerospace and military applications but is now widely used in industrial applications. Electronic timers require no moving parts to generate a time delay. The circuit generally used consists of a resistor–capacitor timing network, a control gate, and a load switching device, which may be either a conventional relay or a reed relay. An electronic time delay relay with a semiconductor load-switching device is called a *timing module*.

Only two types of time delay relays are used much today as electronic components: Motor-driven and electronic.

8.1 DEFINITIONS

ABORTED-CYCLE TIME: Transfer (pickup or dropout) of the switching device which occurs when the timing cycle is aborted or interrupted. Duration can vary from a short pulse to a total time delay, depending on when the interruption occurs in the timing cycle.

ACCURACY (TIMING VARIATION): The difference between the actual time delay obtained and the nominal value specified.

FIRST-TIME EFFECT (ELECTRONIC TIME DELAY RELAYS): The initial timing cycle of electronic time delay relays, particularly those using an electrolytic timing capacitor. The timing cycle may be far out of spec if the timer has not been operated for some time. It is caused by the capacitor not being discharged completely between normal timing cycles but being discharged completely for the first cycle.

RECOVERY TIME: The minimum time required for repeat accuracy within tolerance when a timing cycle is aborted during the timing interval. Recovery time is typically greater than recycle time.

RECYCLE TIME: The minimum length of time between operations that must be allowed to obtain repeat accuracy.

REPEATABILITY: The maximum variation in timing in a group of consecutive timing cycles expressed in percent of a selected reference timing cycle. It is determined at fixed conditions of operating temperature, recycle time and input voltage. (See FIRST-TIME EFFECT)

RESET, AUTOMATIC: The automatic return to the ready state after completion of a timing cycle. Some timing devices must be manually or externally reset.

RESET STATE: The ready condition to which the timer must be returned before timing through another cycle.

RESET TIME: The time required after time-out for the output switch to return to its normal, de-energized state. Do not confuse with recycle time.

SETTING ACCURACY (ADJUSTABLE TIME DELAY RELAYS): The maximum deviation between dial indication and the actual time interval. Setting accuracy is expressed as a percent of the dial indication.

TEMPERATURE ACCURACY: The maximum variation in the time interval measured at any point within a specified temperature range, expressed as a percent of the timing obtained at 25°C.

TIME-OUT: The operational cycle of a time delay relay.

8.2 MOTOR-DRIVEN TIME DELAY RELAYS

A motor-driven time delay relay consists of a timing motor (a motor with an integral gearhead) coupled to an electrically energized clutch, driving a spring-loaded lever through an angular rotation to actuate a switch after a predetermined time interval. (See Fig. 8–5) In this direct

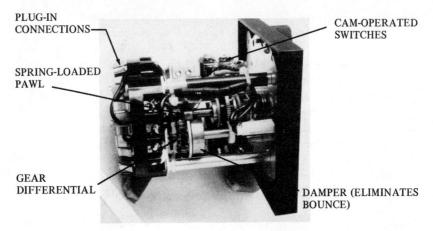

PLUG-IN
CONNECTIONS

CAM-OPERATED
SWITCHES

SPRING-LOADED
PAWL

GEAR
DIFFERENTIAL

DAMPER (ELIMINATES
BOUNCE)

Fig. 8-5 Motor-driven time delay relay. Type shown is Cramer 402. This type of timer has a repeat accuracy of ±0.4%. *(Courtesy of Cramer Division, Conrac Corp.)*

clutch action mode, timing is initiated by energizing both clutch and motor. The load switch is transferred at the end of the time delay and stays in that position until the external control circuit is broken, causing the timer to reset. If the power is interrupted during the timing cycle, the timer resets automatically and will repeat the entire cycle when power is restored provided the external control circuit remains closed.

Motor-driven time delay relays are also available with reverse clutch action. The unit is then normally in a timed-out condition, with the load switch in the normally-open position and the motor circuit open. Closing of the control circuit to the clutch coil disengages the clutching mechanism, which resets the time and closes the circuit to the timing motors. Breaking the control circuit initiates timing. At the end of the cycle, the load and motor switches are transferred and remain in that condition until a new starting impulse is received through the clutch control circuit. If the clutch is momentarily energized during timing, the timer will reset and repeat the cycle. Interruption of power during the time delay causes a suspension of timing but will not reset the timer.

Reverse clutch configuration is used for applications where the timer must not reset with momentary power interruptions during timing. A reverse clutch offers greater circuit flexibility than a direct clutch.

A motor-driven repeat cycle time (Fig. 8-6) consists of a

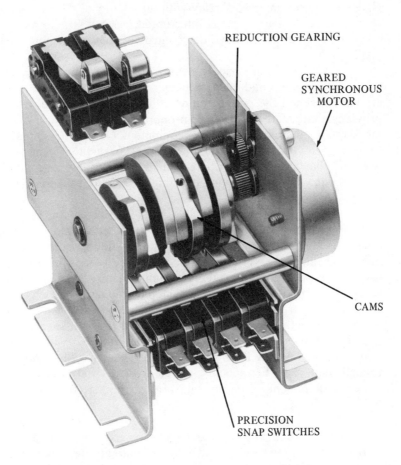

REDUCTION GEARING

GEARED
SYNCHRONOUS
MOTOR

CAMS

PRECISION
SNAP SWITCHES

Fig. 8-6 Motor driven repeat cycle timer. This type of device repeats an ON-OFF timing cycle as long as the control circuit is energized. Complex sequences can be set up with multiple timers. Timer is shown half-size. *(Courtesy of North American Philips Controls Corp.)*

geared-down motor driving a camshaft with cams set to operate switches in a predetermined sequence which repeats continuously. Upon removal of power from the motor, the timer remains in position (i.e., it does not reset), and it starts recycling from this position when power is reapplied. Single-cycle operation can be obtained by wiring the motor through one of the switches to remove power at the end of the cycle.

A proportional delayed-reset motor-driven time delay relay includes the basic time delay relay with an additional differential delay reset feature. When power is interrupted, reset will take place at a constant delayed rate under the control of an escapement mechanism.

These devices are used extensively for delaying application of plate voltage in gas and vacuum power tubes until filaments or heaters have reached the proper temperature. The delayed reset feature provides a circuit reclosure time proportional to the length of power interruption, thereby minimizing equipment OFF time but still affording positive protection for the tubes. These timers can also be used in other applications where similar requirements must be met, such as delaying the restarting of compressors until the starting load has dropped.

Standard models are available with 50, 60, or 400 Hz motors and governed or ungoverned dc motors. In dc applications, the operating voltage range together with required timing accuracy determines whether a governed motor is needed. One-way drives are used in the dc motors to protect the timer against accidental polarity reversal. Up to 10 switches is considered standard; however, timers with more switches can readily be supplied, as can units with inter-camshaft reductions to provide two or more camshaft speeds.

Motor-driven time delay relays can be obtained with open construction; dust-covered for protection against dust, lint, and similar foreign matter; or hermetically-sealed to withstand severe environmental requirements such as salt spray, humidity, fungus, and altitude.

Timing Accuracy. Timing motor speed, the cutting and the setting of the cams, and load switch repeatability determine the timing accuracy of motor-driven timing devices.

The choice of the motor used is therefore extremely important, since it must operate within the speed tolerance over voltage, frequency, and temperature ranges, all of which might cause speed variation. For dc operation, standard dc motors are normally used where cycling time accuracies within ±10% at rated voltage and temperature are required. Governed dc motors are used where repeatability must be held to closer tolerances over wide voltage variations. For ac operation, synchronous motors are used; the accuracy of the motor speed is then dependent upon only the line frequency.

Cams can normally be cut and set for a required sequence with an angular accuracy of ±1 degree of camshaft rotation for single-drop, and ±2 degrees for multiple-drop cams. The effect of this angular tolerance on overall timing accuracy depends on the angle through which the camshaft rotates during the timing interval. Load switch repeatability usually has little effect on overall accuracy.

"Industrial" Motor-Driven Time Delay Relays. These devices (Fig. 8–7) are made in a wide variety of open-frame and dust-cover enclosed models. All have knobs and dials for adjusting the time delay. They are available with 50 or 60 Hz, 115 vac or 230 vac synchronous motors. Each usually has one cam-actuated switch. Internal wiring connections are accessible; the units can be wired or rewired to operate in several modes, and they can be easily interconnected to handle complex timed control sequences. The performance characteristics of typical designs are shown in Table 8–2.

Fig. 8–7 Open-frame industrial time delay relay. Cramer type 360 shown is for "built-in" installations where adjustment of the timing cycle is infrequent and the cabinet can provide physical protection for the unit. Timer is shown half-size. *(Courtesy of Cramer Division, Conrac Corp.)*

Hermetically-Sealed "Military" Motor-Driven Time Delay Relays. These devices are made in standard and subminiature sizes; the subminiature (Fig. 8–8) has one third the volume of the standard size. Models are made with 50, 60, or 4000 Hz synchronous motors and governed or ungoverned dc motors. Two types of governed motors are used, depending on vibration and shock requirements.

The hermetically-sealed housings permit exposure to humidity, salt spray, dust, corrosive or dusty atmospheres, high altitude, and immersion in liquids. Dimensions of a particular unit are determined

TABLE 8-2 Typical Industrial Motor-Driven Time Delay Relays

		Cramer 474/475 Series	Cramer 360 Series	Cramer 402/403 Series
Construction		Panel-mounted, enclosed	Surface-mounted, open frame	Plug-in, enclosed
Operating Voltage		115, 230 vac 50, 60 Hz	115, 230 vac 50, 60 Hz	115, 230 vac 50, 60 Hz
Power Consumption	Motor	2.7 w	4 w	3 w
	Clutch	4.0 w	0 w (slip clutch)	5 w
Operating Temperature Range		−20°C to +55°C	—	−20°C to +55°C
Time Delay Range		0.4 sec to 24 hrs.	5.5, 15, or 60 sec; 5.25 min	0.1 sec to 24 hrs.
Range, single unit Dial division		15 to 1, typical $\frac{1}{15}$th to $\frac{1}{60}$th of full scale	— $\frac{1}{30}$th of full scale	60 to 1, typical $\frac{1}{60}$th of full scale, typical
Time Delay Tolerance (excluding operator error)		±0.6% of full scale (motor running continuously)	±5% full scale	±0.4% of full scale (motor and clutch energized simultaneously)
Repeatability (% full scale)		±0.6% (motor running continuously)	±0.75%	0.4% (motor energized with clutch)
Reset Time		0.5 sec	—	0.5 sec
Switch Rating, Standard		10 A resistive	10 A resistive	10 A resistive
Cost: Each, small qty.		$35	$18	$50

Fig. 8-8 Subminiature hermetically-sealed "military" motor-driven time delay relay. Models are made with 50, 60 and 400 Hz synchronous motors, and governed and ungoverned dc motors. Complex timing functions can be combined in tamper-proof sealed units. Unit is shown half-size. *(Courtesy of North American Philips Controls Corp.)*

by the shape, number, and rating of the switches used, type of motor used, and requirements for RFI filtering.

These relays are all custom-made. Up to seven cam and switch combinations can be ordered in a single unit. Time delay settings for individual switches can be set within a range of 10% to 100% of the overall delay interval. Connections are made to solder hooks set in a glass header; integral connectors are also available.

An advantage of these time delay relays, besides their ability to withstand extreme environmental conditions, is that complex timed control sequences can be obtained in a single tamper-proof package. (Adjustable units are also available.) Standard size and subminiature hermetically-sealed motor-driven time delay relays are compared in Table 8-3.

TABLE 8-3   Hermetically-Sealed Motor-Driven
Time Delay Relays—Summary Characteristics
(North American Philips Controls Corp.)

	Standard	*Subminiature*
Operating Voltages		
Ungoverned dc units†	6, 12, 24, 27.5 vdc	6, 12, 24, 28 vdc
Governed dc units	6, 12, 25 vdc	—
50 or 60 Hz ac units	6, 12, 24, 115, 230 vac	—
400 Hz ac units	115 vac	115 vac
Time Delay Range:		
dc	1 sec to 4 hrs.	25 sec to 3 hrs.
50 or 60 Hz ac	1 sec to 70 hrs.	—
400 Hz ac	1 sec to 6 hrs.	1 sec to 3 hrs.
Time Delay Tolerance		
At 25°C: Ungoverned dc	±10%	±10%
All other	±5%	±5%
Over −54°C to +85°C:		
Ungoverned dc	††	±15%
All other	††	±10%
Repeatability: Best	±3%	††
Load Switch and Rating	SPDT, 10 A resistive	SPDT, 5 A resistive

†To maintain timing accuracy, supply voltage must be regulated.
††Consult manufacturer.

8.3 ELECTRONIC TIME DELAY RELAYS

Most electronic time delay relays (Fig. 8–9) utilize resistor–capacitor networks as a time-generating source (Fig. 8–10). When the switch is closed, capacitor C charges through the resistor R. As the charge on the capacitor increases, the voltage across the capacitor V_c approaches that of the charging source.

The amount of time required for the capacitor to charge through the resistor to a value approximately 63% of the charging source is called the *time constant*. It equals the product of the resistance in ohms and the capacitance in farads.

Increasing the value of either the resistor or the capacitor increases the time constant directly, assuming constant charging voltage. With this basic circuit, time delays ranging from 0.0001 second to 300 seconds are possible, with accuracies as great as 0.5%, depending on the stability of the components.

(a)   (b)   (c)   (d)   (e)

Fig. 8–9 Electronic time delay relays. **(a)** Potter & Brumfield R12/R13/R14/R15 series, **(b)** Potter & Brumfield CD series, **(c)** Potter & Brumfield CJ series, **(d)** Tempo T6, **(e)** Cramer 3800. All units are shown half-size. *(Courtesy of Potter & Brumfield Division, AMF, Inc.; Tempo Instrument, Inc.; Cramer Division, Conrac Corp.)*

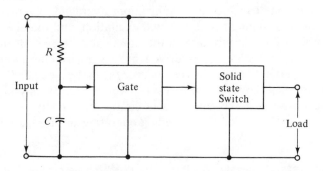

Fig. 8–10   R-C timing circuit.

The gate is biased in the OFF state during the timing cycle. As the capacitor charges, the voltage across it increases exponentially; when the voltage across the capacitor equals the bias voltage, the gate fires, energizing the load. Once fired, the gate locks in the ON state. Most RC timers use a unijunction, or programmable unijunction for a trigger.

Timing Accuracy. Variations in input voltage, or transients during the timing cycle, will affect the timing accuracy by modifying the charging rate of the timing capacitor, or changing the bias on the gate.

Noise or ripple in the supply voltage during the timing cycle may affect the delay time. Negative spikes will momentarily reduce the bias on the gate. If the capacitor has charged sufficiently so that the bias is momentarily less than the capacitor voltage, the gate will fire. These voltage depressions usually shorten the delay time. Early in the charging cycle, negative spikes are more likely to increase the time delay by momentarily reducing the capacitor charging rate. Positive spikes increase the charging rate and reduce the time delay, or the circuit may fire prematurely on the trailing edge of the spike. If the voltage is high and then drops suddenly, the gate may fire instantly, or the delay time will be shortened. Conversely, a step function increase in voltage will lengthen the delay time.

Additionally, several components in the timing circuit are polarity-sensitive, and reversal of the input voltage polarity, when making connections, will cause damage. Transient spikes and overvoltage can also catastrophically damage the timer. Many electronic time delay relays incorporate transient and reverse–polarity protection and voltage regulating circuits.

For variable time delay relays, timing accuracy is usually specified as a percent of full-scale timing under standardized conditions. These conditions can include stabilization, averaging, averaging except for the first timing cycle, nominal excitation, and the like.

Ambient temperature affects the timing accuracy of all time delay relays except those types driven by a synchronous motor and oscillator-counter types. Timing accuracy can be specified at one temperature or a range of temperatures, which may or may not have the same operating temperature range as the time delay relay. Calibration can be destroyed by exceeding operating or storage temperatures.

Oscillator-Counter Timing. In this timing circuit configuration (Fig. 8–11), the time delay is provided by a C/MOS integrated

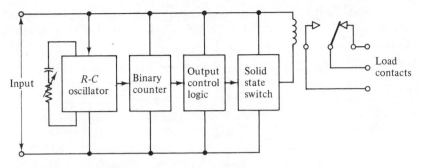

Fig. 8-11   Oscillator-counter timing circuit.

circuit consisting of an RC oscillator, a binary counter, and output control logic. The delay time is set by changing the oscillator frequency by knob adjustment of an internal potentiometer. Standard models provide delays from 0.1 sec to 100 min; delays to one year or longer can be obtained on special order. The delay times provided by an oscillator-binary-counter circuit are not affected by most transients.

Timing Modules.   Electronic time delay relays are sometimes supplied as timing modules (Fig. 8-12) to be used with customer-supplied load-switching relays or to interface with solid-state circuitry. Timing modules have advantages. The timing module can be mated with a relay(s) that has contacts rated for particular loads, avoiding having to use a complete time delay relay to drive other relays just to get required contact ratings. The timing module can be mounted on a PC board, with the load relay out where the load is. If a failure occurs, only the failed part has to be replaced. One timing module could be expected to outlive several relays.

Fig. 8-12   Timing module. An electronic time delay relay with a semiconductor load switch. Potter & Brumfield JT series shown full size. *(Courtesy of Potter & Brumfield, AMF, Inc.)*

Load Switch. A solid-state load switch will provide better operational life and reliability than will an electromagnetic relay, because:

1. The solid-state device has no moving parts;
2. Solid-state switches are normally unaffected by environmental conditions including shock and vibration; and
3. One relay with multiple contact sets can switch more than one independent circuit.
4. Solid-state output devices switch at microsecond rates, while relays require milliseconds. For accuracy in millisecond timing ranges, the solid-state device must be used.

TABLE 8-4   Electronic Time Delay Relays and Timing Modules — Typical Characteristics

	Time Delay Relays		
Class	Industrial	Industrial	Industrial
Type	Cramer 3810/3811	Tempo T-6	Potter & Brumfield CG
Circuit	RC	RC	Oscillator-Counter
Construction	Open frame	Dust cover	Dust cover
	Surface mounted	Octal plug	Octal plug
	Solder tabs		
Fixed Time Delays	None	None	1 sec to 60 min.
Tolerance	−	−	±5%
Adjustable Time Delays	1 sec to 5 min	0.15 sec to 60 sec	0.1 sec to 100 sec
Setting Accuracy	−	±5%	±5%
Range, Single Unit	10:1	10:1	10:1
Repeatability	±4%	±2%	±0.1%
Release Time	−	−	30 ms
Reset Time	−	50 ms	−
Recycle Time	150 ms	100 ms	40 ms
Operating Voltage	115 vac	115 vac	24, 120, 240 vac
			12, 24, 48, 110 vdc
Power Consumption	4 w	5 w	3 w
Load Switching Device	Relay	Relay	Relay
	DPDT, 10 A	DPDT	DPDT, 10 A
Operating Temperature	±10°C to +50°C	−	−10°C to +55°C
Cost: Each, Small Quantity	$24–$36	$30	$60

A solid-state switch, however, does not provide a mechanical disconnect. A small leakage current in the microampere-to-milliampere range will still flow through the load when a solid-state device is switched OFF. The magnitude of the leakage current is determined by the type of device and the conditions under which it is operated.

Operating a relay beyond its rated capacity will shorten contact life. Operating a solid-state switch at excessive levels will cause catastrophic failure.

Table 8–4 compares the characteristics of electronic time delay relays and timing modules.

TABLE 8-4 *Continued*

	Timing Modules	
Industrial	Military	Industrial
Potter & Brumfield CJ	Tempo XTS1/XTS2	Potter & Brumfield JT
RC	RC	RC
Steel case	Crystal can	Plastic encapsulated
Encapsulated	Half crystal can	Low profile
PC pins	Hermetically sealed	Dual thin-line
	PC pins	PC pins
1 sec to 120 sec	25 ms to 180 sec	50 ms to 120 sec
±5%	±10%	±5%
0.1 sec to 120 sec	—	50 ms to 120 sec
±10%	—	±10%
up to 120:1	—	—
±2%	±10%	±2%
50 ms	—	50 ms
—	—	—
50 ms	10 ms	50 ms
12, 24, 28 vdc	28 vdc	12, 24, 48 vdc
—	3 w	—
Dry Reed Relay	Semiconductor	Semiconductor
Various combinations	SPST, 100 ma	SPST, 250 ms
−10°C to +55°C	−55°C to +125°C	−10°C to +55°C
$26	—	$12

8.4 WHO MAKES WHAT

Table 8–5 lists some of the major time delay relay manufacturers.

TABLE 8-5   Time Delay Relay Manufacturers

	(A. W. Haydon) North American Philips	(Cramer) Conrac Corp.	Magnecraft	Ohmite	Potter & Brumfield	Sigma	Struthers–Dunn	Tempo Instrument
Motor-Driven								
Panel-Mounted	X	X						
Open-Frame	X	X					X	
Plug-In		X						
Hermetically Sealed	X	X						
Electronic (RC)								
Panel Mount			X					X
Octal Socket		X	X	X	X		X	X
Rectangular Relay								
Socket/Solder Tabs			X		X	X	X	
PC Board			X		X			
Hermetically Sealed	X							X
Electronic (Oscillator-Counter)								
Octal Socket					X			
Timing Module								
Crystal Can	X							X
PC Flat Pack								X

SUGGESTED READINGS

Grossman, Morris, "Focus on Time Delay Relays." *Electronic Design 21,* October 12, 1975.

Transistorized Industrial Timer Handbook. Tempo Instrument, Inc., Plainview, N. Y., 1965.

CATALOGS

Cramer Division, Conrac Corporation, Old Saybrook, Conn.
Magnecraft Electric Company, Chicago, Ill.
North American Philips Controls Corp., Waterbury, Conn.
Potter & Brumfield Division AMF Inc., Princeton, Ind.
Sigma Instruments, Inc., Braintree, Mass.
Struthers–Dunn, Inc., Pitman, N. J.

CHAPTER NINE

Switches

An electrical switch (Fig. 9–1) is a device for opening and closing an electrical circuit or for rerouting an electrical signal through a circuit. Most commonly, switches are manually operated devices. Toggle, rotary, slide, snap, and rocker switches are widely used. The old-fashioned knife switch is not seen much anymore outside Frankenstein movies. Switches also may be electric-motor driven or solenoid-driven, which makes them *electromagnetic switches* — the same term that is used to describe relays. Unlike relay contacts which however complex have only two positions (open and closed), the contacts of an electromagnetically driven switch have multiple positions; the contact arm is driven, usually by ratchet, to the selected position or through the selected sequence of positions.

Dry reed capsules (see Section 7.9) are also used as switches, with actuation by a moving permanent magnet. These could be better classified as sensors than switches, however. Other devices used as switches include vacuum tubes, neon lamps, incandescent lamps, and semiconductors of many varieties.

Solid-state switches have very long life and high operating speed. They also can handle complex logic economically and in small volume, withstand shock and vibration, operate silently, and generate little radio-frequency interference. But to get all these advantages, circuitry becomes complex and critical. Mechanical switches, on the other hand, are cheap, can survive transients, don't require heat-sinks for high-power loads, can operate at high and low temperatures, can be procured with multiple-pole/multiple-position switching capability in a single package, and have ON resistance in

Fig. 9–1 Switches.

the milliohm rather than the ohm range. OFF resistance is in high megohms, and mechanical switches don't need a power supply for operation. In many instances, the audible click of their operation is a desirable indication of circuit transfer or condition.

The scope of this chapter will be limited to two types of switches: Mechanically actuated switches commonly used with low-power electronic equipment, and stepping switches. The trends in switches for electronic equipment are miniaturization and PC board terminals on both standard-size and miniaturized varieties. There are tiny toggles, push-buttons, slides, rockers, and rotaries, some mounted in TO–5 cases and DIPs (Fig. 9–2). Direct PC board connection of all sizes of switches not only saves installation costs but eliminates a lot of potential wiring errors.

When there is voltage across open switch (or relay) contacts, the switch becomes a capacitor. The capacitance is small and, for most applications, is of no concern. However, switch capacitance can be an important parameter in some audio, video, and high-frequency circuits, just as important as lead-wire capacitance.

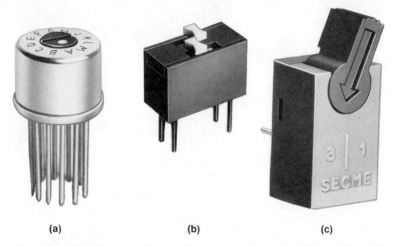

(a)   (b)   (c)

Fig. 9–2 Microminiature switches. These switches are designed for PC board mounting. **(a)** One pole 12 position rotary switch, **(b)** two SPST slide switches in common housing, **(c)** SPDT snap switch. *(Courtesy of Centralab Division Globe-Union, Inc.)*

9.1 DEFINITIONS

Many switch definitions are peculiar to a particular type of switch. The following terms are applicable to all switches except as limited.

ACTUATOR: The mechanical link that drives the plunger of a snap switch.

ADJUSTABLE STOPS: Mechanical stops on a rotary switch that may be adjusted to provide any number of positions desired up to the maximum available on the switch. Adjustment does not require disassembly of the switch.

AUXILIARY ACTUATOR: A mechanism sold separately for attachment to a snap switch. It provides either an easier means of operation and adjustment or an ability to adapt switches to different operating motions.

BIFURCATED CONTACT: A movable or stationary contact that is forked or divided to provide two pairs of mating contact surfaces in parallel for reliability.

COMMON CONTACT, COLLECTOR, OR POLE TERMINAL: In a rotary switch, the contact that is connected to the rotor regardless of rotor position.

CONTACT ARRANGEMENT: The circuitry of a switch.

CONTACT RESISTANCE: The resistance of a pair of closed switch contacts measured from terminal to terminal. It includes the resistance of the contact members themselves.

CONTACT–VOLTAGE DROP: The voltage (which may vary with current magnitude) measured across a pair of closed switch contacts.

CONTINUOUS ROTATION: A rotary switch supplied without stops so that it may be rotated a full 360 degrees in either direction. The switch may or may not have a detent.

CYCLE OF OPERATION: Rotation of a switch from one stop to the other stop and return. In switches without stops—rotation from the first contact to the last and return to the first.

DECK: A section of a rotary switch, sometimes called a *wafer*.

DETENT: The device in a rotary, slide, rocker, or toggle switch mechanism that provides a positive centering of the moving contact relative to the fixed contact at each switch position. It also gives the switch operator a physical and sometimes audible indication that the switch is in the desired position.

DIFFERENTIAL TRAVEL: The distance from the operating point to the release point of a snap switch.

DOUBLE-BREAK CONTACT: A contact combination in which pairs of contacts on a single conductive support simultaneously *open* an electrical circuit connected to the two independent contacts, thus providing two contact air gaps in series when the contact is open.

DOUBLE-MAKE CONTACTS: A contact combination in which pairs of contacts on a single conductive support simultaneously *close* an electrical circuit connected to the two independent contacts, thus providing two contact air gaps in series when the contact is open.

ENCLOSED SWITCH: A switch having all operating parts enclosed by some type of housing.

FIXED STOPS: Mechanical devices incorporated in a rotary switch to limit the rotation of the switch to a predetermined number of positions.

FORCE DIFFERENTIAL: The difference between the plunger operating force and the plunger release force of a snap switch.

INDEXING: The angular displacement in degrees of the movable contact of a rotary switch in moving from one position to the next consecutive position.

LIFE: The number of cycles of operation of a switch during which the electrical and mechanical performance will meet predetermined and stated life-limiting criteria, which may or may not have much meaning in your application.

Nonshorting Contacts (Break-Before-Make): Contacts that, in switching from one position to the next, fully open the first contact circuit before closing the circuit to the next contact.

Open-Frame Switches: Switches without a housing over the contacts.

Operating Force: The force that must be applied to the plunger of a snap switch to cause the moving contact to move from the normal contact position to the operated contact position.

Operating Point: The position of the plunger of a snap switch at which the contacts move from the normal contact position to the operated contact position.

Overtravel: The distance the plunger of a snap switch or push-button switch can be driven past the operating point.

Overtravel Force: The force required to depress the plunger of a push-button or snap switch to the full overtravel point.

Overtravel Point: The position of the plunger of a push-button or snap switch beyond which further overtravel would cause damage to the switch or actuator.

Plunger Free Position: The position of the plunger of a snap switch when no external force is applied to it other than gravity.

Pretravel: The distance through which the plunger of a push-button or snap switch moves when depressed from the free position to the operating point.

Release Force: The level to which force on the plunger of a snap switch must be reduced to allow the contacts to move from the operated contact position to the normal contact position.

Release Point: The position of the plunger of a snap switch at which the contacts move from the operated contact position to the normal contact position.

Release Travel: The distance through which the plunger moves when it returns from the release point to the free position.

Rotational Torque: The force required to move a rotary switch from one position to the next.

Shorting Contacts (Make-Before-Break): Contacts that, in switching from one position to the next, close the circuit to the second contact before opening the circuit to the first.

Spring Return: A spring-loaded device in a switch mechanism that returns the switch to a predetermined position when the operating knob is released.

Stop Strength: The rotational torque that can be applied to the shaft of the switch against the mechanical stops without breaking the stop.

UNIDIRECTIONAL ROTATION: A rotary switch that may be rotated in one direction only.

9.2 PRECISION SNAP SWITCHES

A precision snap-acting switch is a mechanically operated electric switch having predetermined and accurately controlled characteristics. The switch consists of a basic switch used alone, a basic switch used with an actuator, or a basic switch used with an actuator and an enclosure (Fig. 9–3).

Fig. 9–3 Microminiature snap switches. Switches shown half-size. *(Courtesy of Micro Switch Division, Honeywell)*

The moving contact arm of a snap switch is made of spring brass or spring phosphor-bronze strip, and it is so shaped, crimped, and prestressed by heat treatment that when it is anchored inside a snap switch housing, it has two positions of equilibrium (Fig. 9–4). A small deflection of a part of the contact arm by the switch-actuating plunger will cause the contact arm to move rapidly from one position to the other, where it will stay until the plunger is released. When the contact arm moves, it carries the moving-arm contacts from their position against the normally-closed contacts to make contact with the normally-open contacts. Although the actuating plunger movement that produces switch actuation can be as little as 0.001 in., it is normally designed to be nearer 0.03 in.

The same switch is able to control a 15-ampere, 240-volt load,

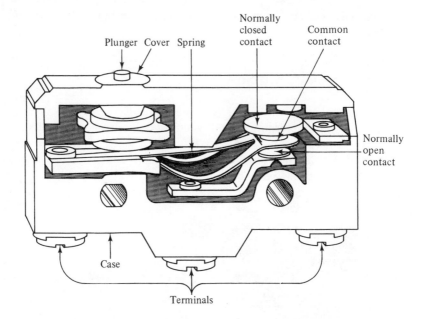

Fig. 9-4 Precision snap switch construction. *(Courtesy of Micro Switch Division, Honeywell)*

or the millivolt–milliampere circuit. Mechanical and electrical characteristics are uniform and stable. The versatility and small size of precision snap switches have led to their use in an endless variety of equipment.

Operation. Figure 9–5 shows a typical precision snap-acting switch. The plunger is in the completely released *free position*, with the normally-closed contacts closed. As the plunger is depressed, it reaches the *operating point*. The distance traveled from the free position to the operating point is called the *pretravel*. When the plunger is depressed to the operating point *without any further movement of the plunger*, the moving contact accelerates away from the normally closed contact, and strikes, bounces, and comes back to rest against the normally-open contact. Movement from the normally-closed contact to the normally-open contact takes place in a few milliseconds. Because of the snap action, the moving contact arm cannot be stopped midway between the normally-closed and normally-open contacts, nor can you do anything to make it move faster or slower.

As the plunger is depressed beyond the operating point, no fur-

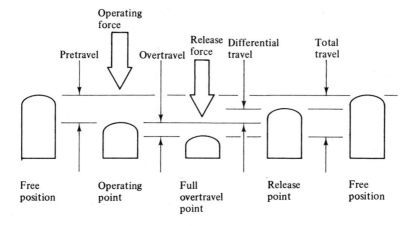

Fig. 9-5 Precision snap switch operation. *(Courtesy of Micro Switch Division, Honeywell)*

ther change in the switch contacts takes place. The distance the plunger can travel beyond the operate point is called *overtravel;* the total distance the plunger moves from the free position to the mechanical stop at the end of overtravel is called the *total travel.*

As the plunger is released, it travels back to the operating point, but nothing happens. The position of the moving contact does not change. As the plunger continues to be released, it reaches the *release point.* At the release point, without any further movement of the plunger, the moving contact accelerates away from the normally open contacts. Once more, within a few milliseconds, the common contact strikes, bounces, and comes to rest against the normally-closed contact, remaining in this position as the plunger is returned to its free position. The distance between the operate and release points is called the *differential travel.* The distance from the release point to the free point is called the *release travel.*

The force required to depress the plunger to the operate point is called the *operating force.* The force required to *hold* the plunger at the release point is called the *release force.* The difference between the operating and release force is called the differential force, or *force differential.*

Basic snap switches are made in a wide variety of physical configurations, sizes, and contact ratings (See Table 9-1). All of the operating parameters can be varied to obtain desired performance characteristics. Snap switch contact forms are shown in Fig. 9-6.

TABLE 9-1 Snap-Action Switches[†]

	Standard	Small	Small, Light Operating Force	Subminiature	Miniature
Size (length)	2 in.	1.25 in.	1.25 in.	0.5 in.	0.75 in.
Contacts	SPST, SPDT			SPDT	
Rating	15 A to 480 vac 0.5 A @ 125 vac	10 A to 250 vac —	3 A to 250 vac 3 A @ 28 vdc	5 A res., 3 A ind.	5 A to 240 vac
Operating Force (typ.)	9 to 13 oz.	6 oz.	1 oz.	5 oz.	3 to 5 oz.
Release (typ.)	4 oz.	1.5 oz.	0.5 oz.	1 oz.	1 oz.
Pretravel	0.015 in.	0.030 in.	0.04 in.	0.020 in.	0.02 in.
Overtravel	0.005 in.	0.025 in.	0.05 in.	0.004 in.	0.01 in.
Cost: Each, small quantity	$2.50	$1.75	$1.50	$3.50	$1.25

† Data given for general-purpose versions.

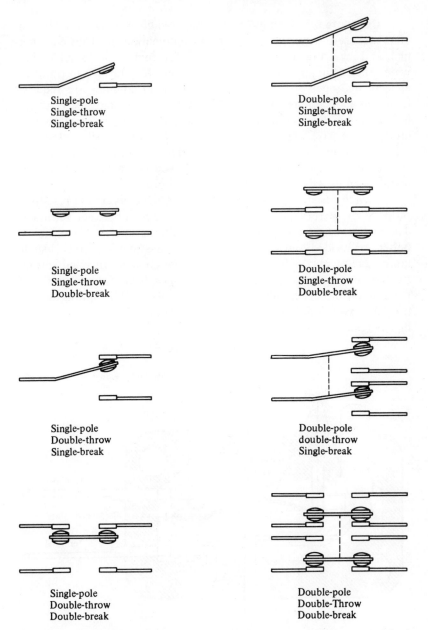

Fig. 9-6 Precision snap switch contact forms. *(Courtesy of Micro Switch Division, Honeywell)*

Contacts. Generally, if the snap switch is sealed, silver contacts are used. Silver offers the best combination of electrical, thermal, and mechanical properties needed for contact performance in a wide range of applications. Silver, however, does tarnish in the presence of hydrogen sulfide and water, and gold contacts are required for some applications. If the snap switch is not sealed, the electrical load must be considered. Silver contacts are required if there will be arcing; otherwise, the switch environment governs. Bifurcated contacts should be used if dust and other particle contaminants can get to the contacts; gold will not help. Alloyed contacts, using various metals, can usually be obtained on special order.

Switch resistance (contact resistance) depends on the resistivity of all of the parts making up the conducting path between terminals: The resistance across the contact faces; the resistance across staked, bolted, or welded joints; the resistance across moving joints, usually knife-edge pivots; and the resistivity of the contact arms and springs. Switch resistance for snap switches ranges from 5 to 50 mΩ, depending on the design and materials (Fig. 9–7). (The spring material used in a switch designed for high temperature operation would have higher resistivity.)

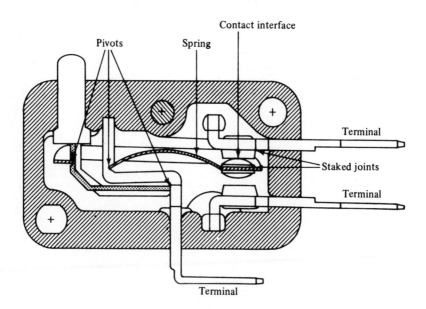

Fig. 9–7 Sources of resistance in precision snap switches. *(Courtesy of Micro Switch Division, Honeywell)*

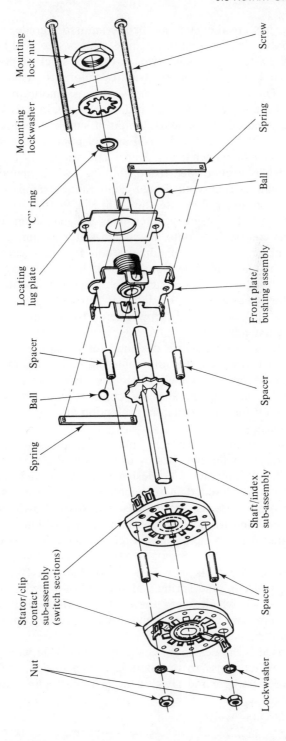

Fig. 9-9 Rotary switch construction.
(Courtesy of Centralab Electronics Division Globe-Union, Inc.)

rials and finishes are the same as for the industrial grade; for a 200-hour salt spray, nickel-plated brass and stainless steel will be used.

Rotary switches are sold two ways: As stock items through distributors and as special-order switches tailored to your exact requirements. The economics of rotary switch assembly are such that it doesn't take a very big special order to result in a cost savings over buying standard components. To put it simply, there is no point in paying for contacts and decks you are not going to use when you can readily get exactly what you want. The variations possible are endless, and it is a good idea to consult the switch manufacturers' engineering staff early in your design cycle. Some examples of rotary switch contact configurations are shown in Fig. 9–10.

Switch diameter basically determines the maximum number of positions that can be fitted onto the switch stator assuming the same materials and general style of construction for the switches being compared. The angle of throw between switch contact positions also affects the number of positions, and each size switch wafer may be fabricated for an assortment of different throws. Table 9–2 shows combinations of diameter and throw offered by one manufacturer in standard open switches.

Some rotary switches are designed for mounting directly on a printed circuit board by either dip soldering or hand soldering. Figure 9–11 shows one such switch. Rotary switches are also made with the index mechanism and the contacts completely enclosed (Fig. 9–12) both for environmental protection and to reduce production line damage.

Miniature rotary switches are made with the same number of poles as larger-diameter switches by using molded stators and rotors, usually of diallyl phthalate. Clip design is changed, and barriers between contacts are molded in to maintain voltage ratings. Molded stators and rotors are also used on standard-sized switches to improve reliability and to increase the number of positions that can be accommodated without increasing switch diameter (Fig. 9–13).

Contacts. Contacts in rotary switches are of two basic designs: Blade and clip, and spring and stud. Both wipe as they make and break contact, which keeps the contact surface clean of oxides and organic films and improves contact resistance by continually exposing fresh metal surfaces.

Most used is the design in which the rotor blade wipes between stationary contact clips. In this double-wiping design, the blade is forced between compressed jaws of the stationary clip, making con-

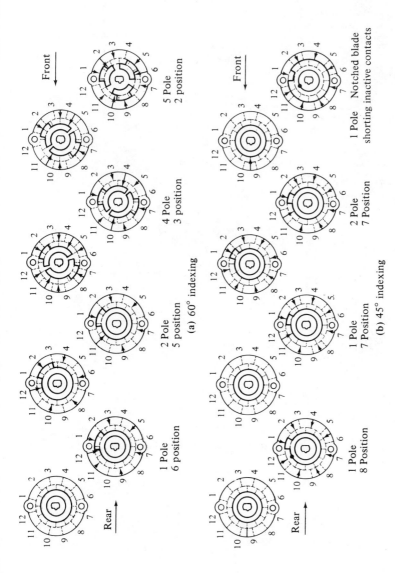

Fig. 9-10 Typical rotary switch contact configurations. *(Courtesy of Oak Industries, Inc.)*

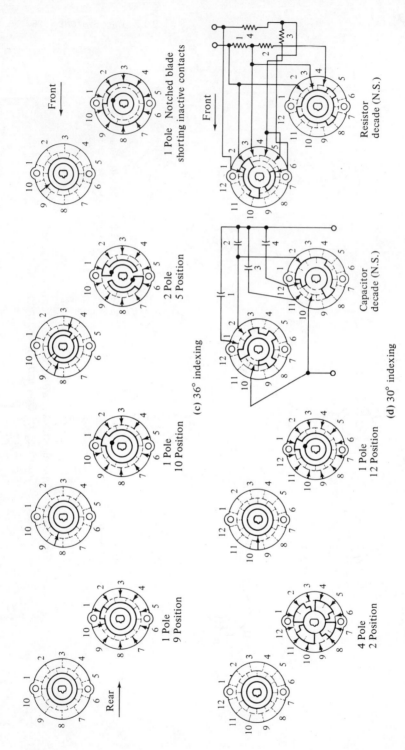

(c) 36° indexing

1 Pole   Notched blade
shorting inactive contacts

2 Pole
5 Position

1 Pole
10 Position

1 Pole
9 Position

Front

Rear

Resistor
decade (N.S.)

Capacitor
decade (N.S.)

(d) 30° indexing

1 Pole
12 Position

4 Pole
2 Position

Front

Fig. 9-10 *continued*

Table 9-2   Rotary Switches†

Diameter (inches)	Throw and Positions										
	90° (4)	60° (6)	45° (8)	40° (9)	36° (10)	30° (12)	25.7° (14)	22.5° (16)	20° (18)	18° (20)	15° (24)
1		X	X		X	X					
$1\frac{5}{16}$	X	X	X		X	X					
$1\frac{1}{2}$	X	X	X		X	X	X				
$1\frac{7}{8}$	X	X	X	X	X	X		x	X	X	
$2\frac{5}{16}$				X		X			X		X
$2\frac{13}{16}$								X			

†Throw and positions available on standard open-construction switches.

Fig. 9–11 Rotary switch for PC board mounting. Oak 24 position switch can be either hand or dip soldered to PC board. A variety of contact materials can be specified. Switch is shown two-thirds size. *(Courtesy of Oak Industries, Inc.)*

tact with both sides of the blade. The sides of the jaws are flared to guide and center the blade.

For high-current or low-contact resistance applications, the multiple-spring and stud design is used. Typical uses are in decades, bridges, and high-quality audio attenuators. These switches are bulky, and the physical design requirements make them expensive.

Rotary switches come in shorting and nonshorting styles. In the shorting style, the rotor blade is wide enough to bridge adjacent

(a)　　　　　　　　　　　　　　　　**(b)**

Fig. 9–12 Enclosed rotary switches. **(a)** Type 100 miniature, and **(b)** Type 80 subminiature. The type 80 switch can be mounted on a PC board, but only partially connected to the board without jumpers. *(Courtesy of Stackpole Component Co.)*

fixed contacts and momentarily short them together when the switch position is changed from one position to the other. This make-before-break action is desirable in audio applications, for example, where a momentary open circuit introduces clicks and pops and other undesirable transients. In nonshorting switches, a narrow rotor blade breaks one circuit before making the next.

Contact Material. Selection is based mainly on four factors: The required switch life; the characteristics of the load circuit (whether resistive or inductive); the operating environment (temperature, humidity, atmospheric contaminants); and cost. Choice of material is more critical for low-level and dry-circuit switching than it is for higher currents, because the higher currents can punch through surface films and contaminants. The high cost of all precious metals is leading to more inlaying and selective plating of contact surfaces only and is leading away from precious-metal clips.

Silver-Plated Brass. With 0.0001-in. minimum plating, silver-to-silver contact is provided for 10,000 cycles at no load or 6000 cycles at rated load. Rated load is typically 1 A at 28 vdc or 0.5 A at

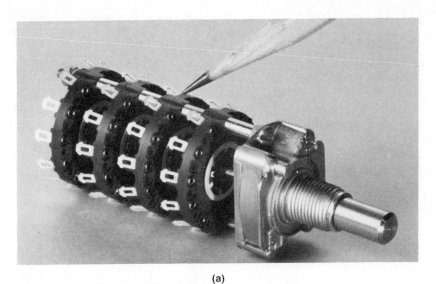

(a)

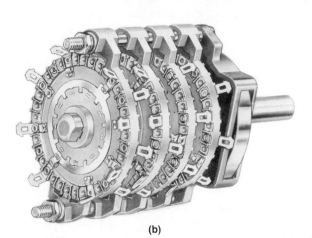

(b)

Fig. 9-13 Molded rotary switches. The stators and rotors of these switches are molded of diallyl phthalate. **(a)** One-inch Acorn®. **(b)** One and a half inch Multidex®. *(Courtesy of Oak Industries, Inc.)*

110 vac. After the plating is worn through, brass-to-brass contact is still made, but contact resistance becomes erratic.

Silver Alloy. Silver-to-silver contact is provided for 200,000 cycles at no load or 50,000 cycles at rated load. For the same configuration contacts as for silver-plated brass, the load rating would typically be 2 A at 28 vdc and 1 A at 110 vac. Contact resistance is uniform over the life of the switch.

Gold-Plated Silver Alloy. The end product of this combination depends on the thickness and type of the gold plating. With 0.000015 in. of gold, performance will be the same as for plain silver alloy except that shelf life is increased and soldering will be easier. With a 0.0002-in. hard gold plating, 50,000 cycles of gold-to-gold contact are provided for dry-circuit applications, plus minimum variation in contact resistance and reliable contact even with infrequent operation or in mildly corrosive atmospheres. After plating is worn through, contact is silver-to-silver for the rest of the life of the switch.

Nickel Alloy, Hard Gold Overlay. This contact material for dry-circuit switching at temperatures up to 150°C and slightly corrosive atmospheres provides gold-to-gold contact for 50,000 to 100,000 cycles. Contact resistance is higher than for gold-plated silver alloy contacts because of the higher resistivity of the contact base material.

Gold Alloy. These contacts are ideal for switches used infrequently, because they do not oxidize the way silver contacts do. With the price of gold today, you had better keep the switch locked up.

Table 9–3 compares practical combinations of rotary switch contact materials.

Insulation.   The choice of switch insulating material is a major factor in the life expectancy of the switch. Reduction of dielectric strength and insulation resistance, both of which deteriorate with switch use, usually determine the end-of-life point of the switch, and the end of life is frequently catastrophic.

Insulation resistance between contacts drops steadily with use because of the buildup of material particles rubbed off the contacts, atmospheric moisture on the insulator surface, and other airborne contaminants. Some switches can be cleaned to prolong life, others cannot because of the application. Actual dielectric breakdown in operation is much less likely.

Some insulation materials are:

Paper Base Phenolic. (XXXP Grade or PEB per MIL S–3786) This economical insulation material is used in commercial-grade applications. It is the most widely used switch insulation. The top temperature is 85°C; minimum insulation resistance is low—only 1000 MΩ. Phenolic is not recommended for high voltage circuitry or for applications requiring high insulation resistance.

Ceramic. Three times more expensive than phenolic, ceramic is stronger, more rigid, has a temperature rating of 150°C and a min-

TABLE 9-3 Rotary Switch Contact Materials

Clip Material	Application	Maximum Temperature	Life	Rotor Blade Material	Rated Load (Typical)
Brass, silver-plated	Commercial	+100°C	10,000 cycles no load; 6,000 cycles rated load	Brass, silver-plated	1 A 28 vdc 0.5 A 110 vdc
Silver alloy	Industrial	+100°C	200,000 cycles no load; 50,000 cycles rated load	Silver alloy	2 A 28 vdc 1 A 110 vdc
Silver alloy: 0.000015 gold plate (min.)	Military	See MIL MIL S–3786	See MIL S–3786	Silver alloy	See MIL S–3786
Nickel alloy: hard gold alloy rolled or contact surface	Dry circuit	150°C	50,000 cycles	Silver alloy: 0.0002 hard gold plate (min.)	—
Silver alloy: 0.0005 gold alloy rolled or contact surface	Dry circuit	100°C	50,000 cycles	Silver alloy, gold overlay	—

imum insulation resistance of 10,000 MΩ. It is usually coated with DC–200 silicone fluid to prevent surface wetting, which would reduce insulation resistance. The high dielectric constant of ceramic results in a high capacitance between contacts (higher than with phenolic or glass silicone insulation), requiring caution in its use in high-frequency circuits.

Epoxy-Glass. Four times more expensive then phenolic, this material has good structural and dielectric strength, with low moisture absorption. Minimum insulation resistance is 10,000 MΩ, and the material can be used to 125°C.

Silicone-Glass. Costing slightly more than epoxy-glass, silicone-glass has lower mechanical strength but better RF characteristics. The insulation resistance and operating temperature are the same.

Diallyl Phthalate. This molded insulation material costs about six times more than phenolic, has excellent structural strength (but is brittle), has low moisture absorption, and is not affected by the solvents and cleaners most commonly used. The top operating temperature is 180°C, and the minimum insulation resistance is 10,000 MΩ. Being a pressure-molded material, it allows greater flexibility in switch design than other materials, particularly in contact placement.

Kel–F. Costing almost ten times more than phenolic, this material is used in critical applications. It has excellent structural strength, minimum insulation resistance is 20,000 MΩ, the top operating temperature is 125°C, and moisture absorption is low.

Mycalex. Costing slightly more than Kel–F, mycalex has good structural strength, the lowest moisture resistance of all listed materials, high dielectric strength, a top operating temperature of 150°C, and an insulation resistance of 10,000 MΩ. Table 9–4 compares characteristics of typical rotary switches.

9.4 LEVER SWITCHES

Lever switches, telephone-type lever switches, or stack switches (Fig. 9–14) are designed for switching audio circuits, and they do it effectively and quietly. Contacts are rated from 1 A to 5 A, 300 watts maximum ac, noninductive. Contacts are not rated for dc, and most problems with noisy audio switching occur because of attempts to switch dc along with audio.

Contacts are arranged in flat spring pileups much the same as in telephone relays, and contact combinations are available in many of the same forms. Depressing the lever handle flexes moving con-

TABLE 9-4 Typical Rotary Switches

| | Open Contacts | | | Enclosed Contacts | | |
	High-Power	Low-Cost	Premium-Quality	Standard	Miniature	Miniature
	Centralab Series 230	Oak J, K	Oak Multidex	Stackpole Series 100	Stackpole Series 80	Oak
Contact Rating (standard)	9 A @ 115 vac 11 A @ 28 vdc	0.5 A @ 110 vac 1 A @ 28 vdc	0.5 A @ 110 vac 1.0 A @ 28 vdc	0.5 A @ 125 vac 1.0 A @ 28 vdc	50 mA @ 125 vac 125 mA @ 28 vdc	250 mA @ 110 vac 500 mA @ 30 vdc
Size (in.)	$2\frac{13}{16}$ dia	$1\frac{17}{32}$ dia	$1\frac{9}{16}$ dia	$1\frac{1}{8}$ dia	$\frac{13}{16}$ long	$\frac{11}{16}$ dia
Positions (max.)	18 (20°)	14 (25.7°)	36 (10°)	18 (15°)	12 (30°)	12 (30°)
Decks, maximum number	4	5	5	40	10	5
Insulation Stator	Ceramic	Phenolic	Diallyl phthalate	Diallyl phthalate	Diallyl phthalate	Diallyl phthalate
Insulation Rotor	Ceramic	Phenolic	Diallyl phthalate	Diallyl phthalate	Delrin	Glass-filled Kel-F
Contacts Stator	Silver alloy buttons	Brass, silver-plated	Silver alloy	Silver	Brass, silver-plated	Silver alloy
Contacts Rotor	Silver alloy	Brass, silver-plated	Silver alloy	Silver	Silver alloy	Silver alloy
Detent	Dual ball	Dual ball	Single or dual ball	Dual ball	Dual ball	Tri-ball
Adjustable Stops	Yes	Yes	—	—	—	—
Index Life	25,000 cycles	6000 cycles	200,000 cycles	—	50,000 cycles	10,000 cycles
Cost: Each, small qty. Two deck switch, max. contacts.	$23	$6	$50	$12	$7	$14

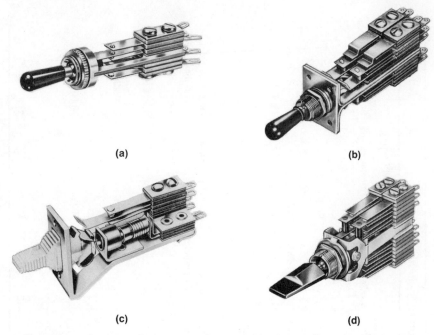

(a)

(b)

(c)

(d)

Fig. 9–14 Lever switches. **(a)** Series 3000 Lev-R ®, **(b)** Series 16000 Telever ®, **(c)** Series 25000 Lever-Lite ®, **(d)** Series 41000 LT switch. Switches are shown half-size. *(Courtesy of Switchcraft, Inc.)*

tact springs, usually with a rolling cam; other moving contact springs in the pileup are actuated by buffers. Contact combinations up to 8PDT are standard. Terminals available include solder and wire-wrap.

These switches have two or three positions. Either or both of the outside positions may be obtained with spring return for momentary or nonlocking contact.

Mounting methods vary. Older types are attached to panels with machine screws, which require either an escutcheon plate on the switch or time-consuming panel machining, or special punches. Newer types have threaded bushings and mount in a single round hole. Turning in the hole is prevented either with a keyed hole or by using a nonturning washer keyed into an adjacent small round hole. One line of illuminated lever switches mounts either in a punched rectangular hole (plus two screw holes) or in three round holes with an escutcheon plate.

Contacts. Fine silver is the standard contact material. Palladium or gold alloy welded crossbar contacts or silver cadmium oxide contacts are also available on special order. Contacts can be easily cleaned to prolong life.

One variety of lever switch is designed to have extremely low intercontact capacitance (Fig. 9–15). Once known as Federal anticapacity switches, these switches have flat actuator springs and round contact springs spaced wide apart in a molded insulating block. Exact capacity cannot be specified, but it is lower than for any other lever switch.

Fig. 9–15 Low capacitance switch. The configuration of the contact springs produces extremely low spring-to-spring capacitance. The switch is used primarily in test instruments. Series 10000 switch is shown half-size. *(Courtesy of Switchcraft, Inc.)*

If you want the advantages of a lever switch in a switch of your own special configuration, you can purchase contact pileups with flat or *V*-formed actuators for actuation by your own cam or pushrod. Fig. 9–16 shows typical pileups.

Table 9–5 compares typical lever switches.

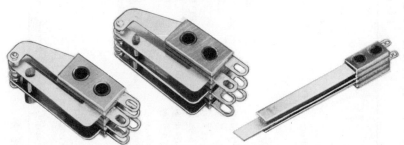

Fig. 9–16 Switch pileup. These assemblies can be purchased separately for building custom switches actuated by cams, push rods, etc. *(Courtesy of Switchcraft, Inc.)*

TABLE 9-5 Lever Switches—Typical Examples (Switchcraft)†

	Lev-R Switch	Lever-Lite III Illuminated Handle	Telever Switch	LT-Switch	Low-Capacity Switch
Lamp	N/A	T–1¾	N/A	N/A	N/A
Mounting	Single round hole	Rectangular hole	Rectangular hole plus screw holes	Single round hole	Rectangular hole plus screw holes
Frame	Brass, nickel-plated	Plastic	Steel, iridite over cadmium	Steel, iridite over cadmium	Aluminum
Knob (standard)	Bat handle, black, red	Paddle	Bat handle, black, red	Paddle	Bat handle, black
Contact Forms (max.)					
Two position switch	3PDT	8PDT	8PDT	4PDT	2PDT
Three position switch (each side)	2PDT	8PDT	4PDT	4PDT	4PDT
Current Rating (Noninductive load)	3 A, 300 w ac	2 A, 200 w ac	2 A, 200 w ac	3 A, 200 w ac	1 A, 100 w ac
Springs	Nickel silver	Phosphor bronze, silver-plated	Nickel silver	Nickel silver	Tempered phosphor bronze
Contact Material (standard)	Fine silver	Welded crossbar, palladium	Welded crossbar palladium	Welded crossbar, palladium	Fine silver
Cost: Each, small qty.	$4–$7	$11–$19	$8–$16	$6–$15	$10–$26

†All available two or three position, nonlocking and locking.

9.5 SLIDE, TOGGLE, AND ROCKER SWITCHES

All three types of switches are made in a variety of physical sizes, contact forms, load ratings, and grades.

Slide Switch.   Slide switches are made in standard, miniature, and microminiature sizes. Figure 9–17 shows several varieties of standard and miniature slide switches. The basic slide switch consists of a spring-loaded strip copper-alloy moving contact that is slid back and forth by a small knob across stud contacts fixed in a plastic insulator base. Construction varies, but that shown in Fig. 9–18 is typical.

Fig. 9–17   Standard and miniature slide switches. Shown three-quarter size. *(Courtesy of Stackpole Components Co.)*

Contacts are silver-plated and are available in a variety of forms up to three-pole/three-position and double-pole/four-position. Some contact forms are available without detents or with spring return. Refinements include double-wipe contacts and a snap-slide action in which a bifurcated contact is lifted off the fixed contacts as it is moved and then allowed to drop against the fixed contact.

The simple construction of slide switches results in low cost. They are used for many applications in consumer, electronic, automobile, and office equipment, in test equipment, in electronic instrumentation, and in small appliances. Many are tested and qualified by both Underwriters' Laboratories, Inc. and the Canadian Standards Association.

Toggle Switches.   The lever of a basic toggle switch operates a spring-loaded toggle linkage that produces rapid contact transfer.

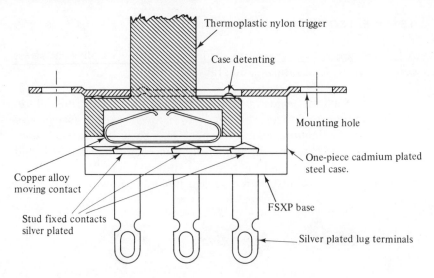

Thermoplastic nylon trigger

Case detenting

Mounting hole

One-piece cadmium plated steel case.

Copper alloy moving contact

FSXP base

Stud fixed contacts silver plated

Silver plated lug terminals

Fig. 9–18 Slide switch construction. *(Courtesy of Stackpole Components Co.)*

The action is not the same as in a precision snap switch, because the moving contact arm does not accelerate away from the actuator. Toggle switches are also manufactured with snap-acting contacts replacing the true toggle mechanism and with toggle-mechanism actuation precision snap switches.

Rocker Switches. A rocker switch is a slide, toggle, or precision snap-acting switch with an attached rocker handle (Fig. 9–19). They are made in several sizes and in a variety of contact forms and current ratings. The plastic handles are available in many colors, some with legends.

There are two varieties of lighted rocker switches: Opaque handles with miniature lamps imbedded in the surface, and translucent handles illuminated by transmitted light from behind the handle, as in a display push-button. The lamp circuit is isolated from the switch.

9.6 PUSH-BUTTON SWITCHES

These switches can be separated into four classes: Plain push-buttons, illuminated push-buttons, illuminated display push-buttons,

TABLE 9-6 Slide Switches—Typical Characteristics

	Standard	Miniature	Microminiature
Enclosure Size (DPDT Switch)	1.25 x 0.5 x 0.5 in.	0.63 x 0.43 x 0.25 in.	0.41 x 0.2 x 0.24 in.
Contact Forms	SPST DPDT 3PDT SPDT DP3T SP3T DP4T	SPST DPST SPDT DPDT SP3T DP3T	SPST (2) DPST DPST (reversing)
Contact Ratings	3 A to 10 A, 115 vac 0.5 A to 1 A, 115 vdc gold-plated contacts available for low- level circuits	1 A to 3 A, 115 vac 0.5 A, 115 vdc ratings are for inductive loads	Not rated for ac 0.3 A @ 24 vdc gold-plated over nickel standard
UL/CSA Listing	Yes†	Yes†	No
Cost: Each, small qty.	30¢–40¢	30¢–60¢	$2.64

† Not all varieties.

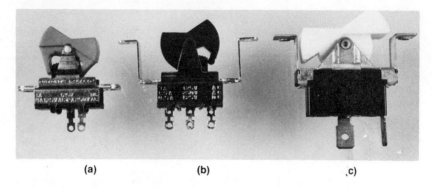

Fig. 9–19 Rocker switches **(a)** and **(b)** Slide-switch mechanism *(Courtesy of Stackpole Component Co.)*. **(c)** Toggle mechanism. *(Courtesy of Cutler-Hammer)*

and mechanically interlocked push-buttons that can be either plain or illuminated. Contacts operate in four modes:

Momentary Positive. Contacts transfer when the button is depressed and transfer back to their original position when the button is released. At transfer, there is usually an audible click and a definite feel to the switch, both of which indicate that the switch has been actuated.

Momentary Silent. Contacts transfer when the button is depressed and transfer back to their original position when the button is released. Contacts transfer silently, and there is no perceptible feel to the switch at actuation, although the operator would feel the button hit bottom at the end of its travel.

Alternate Action. The contacts transfer when the button is depressed and latch in that new position; they do not transfer back to the original position when the button is released. The next time the button is depressed, the contacts do transfer and return to their original position.

Push-Lock/Push-Release. The contacts transfer when the button is depressed the same as in the alternate-action mode; but when the button latches, it remains in a depressed position. Pressing the depressed button releases the latch, and the button and contacts return to their original positions.

Push-button switches are often used in conjunction with relays. The circuit transfer is initiated by a momentary-action push-button and held by relay contacts.

Plain and Illuminated Push-buttons. Individual plain and lighted push-button switches are made in many styles. Contact configurations include toggles, snap switches, and lever-switch style pileups. Contacts may be open or enclosed. A lighted push-button has the advantage of combining control and indicating functions in a single device conserving panel space and reducing operator error. In lighted switches, the control and indicator circuitry can be kept electrically isolated. Switch mounting should be a factor in selection; panel preparation for round hole mounting is easier and cheaper than square or rectangular hole mounting.

Display Push-buttons. These switches have rectangular or square translucent plastic push-buttons (Fig. 9–20). They are designed to be used side-by-side in horizontal or vertical rows for the control and indication of operational sequences of complex equipment. Modular construction permits great functional and operational versatility. It also simplifies maintenance. The switches can also be used individually. Matching indicator assemblies are available.

Display push-buttons are illuminated either by transmitted color or by projected color. Both systems have advantages. *Transmitted color illumination* uses a colored push-button and a clear lamp. This method is used when the color of the push-button must be distinguished when the display is unlighted. *Projected color illumination* uses a white translucent push-button and a colored lamp. The lamp is colored by a colored plastic boot fitted over its end or by a colored filter placed inside the push-button. The push-button is white when unlighted. With projected color, up to four different colors can be displayed in a single display push-button; the screen may be split vertically, horizontally quartered, or quartered in one half only. The characteristics of three brands of display push-button switches are described in Table 9–7.

Switches can be mounted with flexible latches or clips or with flanges, bushings, or clamps, which provide greater rigidity. Replacement of lamps from the front is preferred to rear replacement because other components are often in the way. Some switches have finger-tip lamp replacement; others require a tool.

Mechanically Interlocked Push-buttons. These switches (Fig. 9–21) are used in applications in which circuits must be switched randomly rather than sequentially as circuits are switched by a rotary switch, and positive mechanical interlock is required. Positive mechanical interlock is a requirement of the application, not the switch. Individual push-button switches are arranged in a row on a frame. This interlocking can take several forms depending on desired switching control, but not all interlocking modes are available on all types or on all styles within a type.

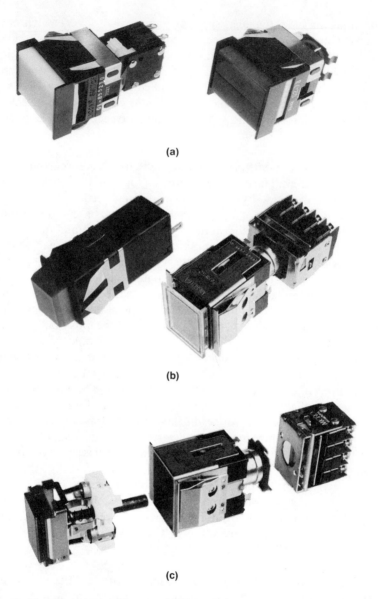

(a)

(b)

(c)

Fig. 9–20 Display pushbutton switches. **(a)** Licon type 01-800 single-lamp indicator/pushbutton switch; indicator module of matching type 02 two-lamp indicator/pushbutton switch. *(Courtesy of Licon Division, Illinois Tool Works, Inc.)* **(b)** Micro Switch Series 4 single lamp indicator/pushbutton switch, Series 2 modular barrier mount indicator/pushbutton switch. **(c)** Disassembled Series 4 switch. *(Courtesy of Micro Switch Division, Honeywell)*

TABLE 9-7   Rectangular/Square Display Modular Lighted Push-Button Switches and Indicators†

	Licon			Micro Switch		Switchcraft
Series	01–800 Series	02 Series	04 Series	Series 4	Series 2	Push-lite
Display Size (in.)	0.87 x 1.12	0.87 x 1.12	0.88 sq.	0.50 x 0.69	0.85 x 1.00	0.74 x 0.83
Screen Division	Full only	Full, H or V split	Full, H or V split, quartered	Full screen only	Full, H or V split, quartered	Full, H or V split
Lamps	1	2	4	1	4	1 or 2
Lamp Style	T-1¾ flange base	T-1¾ flange base	T-1¾ flange base	T-1¾ wedge base	T-1¾ flange base	T-1¾ flange base
Illumination	Transmitted or projected	Transmitted or projected	Transmitted or projected	Transmitted, projected, or hidden color	Transmitted or projected	Transmitted or projected
Mounting	Integral barrier	Integral barrier	Barrier or flange	Integral barrier or bezel	Barrier or flange	Barrier or flange
Mounting Contact	1.14 x 1.00	1.14 x 1.00	1.00 x 0.89	0.98 x 0.79	1.00 x 0.86	1.00 x 0.88
Switch Type and Standard Contacts (other contacts available)	Snap switch SPDT to 6PDT 10A resistive	Snap switch SPDT to 6PDT 10 A resistive	Snap switch SPDT to 6PDT 10 A resistive	Snap switch SPDT, DPDT 5 A	Snap switch SPDT to 4PDT 5 A	Leaf switch SPST to 4PDT 2 A/200 w Snap switch 5 A
Operating Force	18–96 oz.	18–96 oz.	60–96 oz.	Not given	Not given	Not given
Switching Modes	Momentary, alternate	Momentary, alternate, solenoid hold	Momentary, alternate, solenoid hold, solenoid reset	Momentary, alternate	Momentary, alternate, solenoid hold	Momentary, push-lock/ push-release

†All are relamped from the front.

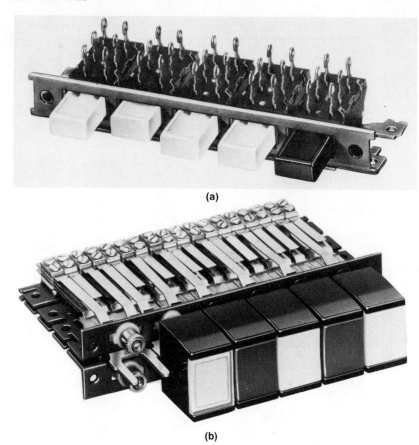

Fig. 9–21 Mechanically-interlocked pushbutton switches. **(a)** Series 6500 low cost, high quality, non-illuminated multiple-station pushbutton switches, can be obtained with up to 18 switches in a single row; **(b)** Series 3800 lighted multiple-station push-button switch, uses miniature stack switches instead of sliding switches. Switches are shown half-size. *(Courtesy of Switchcraft, Inc.)*

Normal Latching (Interlock). Depressing any button releases any previously depressed button. (Two or more buttons may be depressed and their circuits transferred together if the buttons are depressed simultaneously.)

Momentary. Latching is made inoperative on any one or more buttons, but independent spring return action must be substituted.

Accumulative Latching (Interlock). Buttons stay down and do not release when another button is depressed. All are released when a separate release button is depressed.

Notched Latching. Buttons can be arranged to latch independently of any other button.

Push-Push. A button latched independently of other buttons. The button is depressed once to transfer contacts and latch and depresses a second time to release.

Group Latching (Interlock). Two groups of buttons on a push-button switch bank may be set up to have normal latching independently — in other words, two mechanically interlocked push-button switches on a single frame.

Blockout. A blockout mechanism can prevent more than one button latching at a time regardless of how many are depressed, but it normally does not prevent momentary-action contact transfer.

In most mechanically-interlocked push-button switches the contact configuration is similar to that of a slide switch. Miniature stack switches and snap switches are also used in some styles. The switches are made with a variety of push-button shapes, with and without illumination.

Colors and Legends. Six colors are available for switches with illuminated and display push-buttons: Red, amber, green, white, blue, and yellow.

Red is used to indicate an unwanted condition, such as an error, malfunction, failure, etc.

Amber is used as a warning of an unwanted condition.

Green is used to indicate a satisfactory condition, such as ready, complete, normal, etc.

White is used to indicate the status of a function or piece of equipment without any implication of satisfactory or unsatisfactory condition.

Blue is used for advisory messages.

Yellow is sometimes substituted for amber, but the two colors should never be used on the same panel.

Legends and messages can be engraved or hot-stamped on illuminated and display push-buttons. Messages should be kept brief, require no decoding, and use commonly accepted abbreviations such as those in Mil–Std–12.

The layout of lighted push-button switches on a panel should be designed with their functional interrelationship as the prime consideration. Barriers between switches, or *bezels,* should be used to prevent accidental operation of adjacent switches; related units should be grouped. It is important that the panel have a functional as well as pleasing appearance.

9.7 ROTARY STEPPING SWITCH

A rotary stepping switch (Fig. 9–22) rotates, steps, and switches (by comparison, a stepping relay merely switches—opening and closing one set of contacts). The switch has one or more sets of pairs of wiping contacts called *wipers,* which are fixed to a shaft. Each wiper pair is insulated from the shaft and from the other wiper pairs.

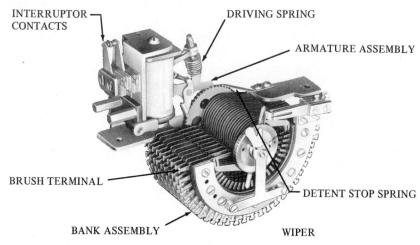

Fig. 9–22 Rotary stepping switch. The type 45 switch shown has ten 26-position bank levels. Switch operation can be either impulse-controlled or self-interrupted. *(Courtesy of GTE Automatic Electric)*

Stationary contacts, insulated from each other and the wiper, are arranged in crescent-shaped levels. There is one level for each set of wipers. The levels are assembled in banks, and each level is insulated from those adjacent to it. There may be up to 400 contacts in a single bank.

Switches are usually indirectly driven. Energizing the stepping electromagnet moves the driving pawl out of the ratchet and drops it over the succeeding tooth but does not move the wiper assembly; when the magnet is de-energized, the wiper assembly is driven forward by a spring. After each pulse to the stepping magnet, the wipers move one step. The switches are capable of running at 65 steps per second.

Several wiper contact configurations are available; rotary stepping switches are normally equipped to transfer with a nonbridging break-before-make contact action. If desired, a bridging make-before-break action can be furnished. Standard wipers and bank contacts are phosphor-bronze. Gold-plated contacts are used for dry-circuit and low-level applications. Characteristics of a typical large and a typical small stepping switch are compared in Table 9–8.

TABLE 9-8   Rotary Stepping Switches†

| | Automatic Electric | |
	Type 45	Type 88
Capacity	Two to twelve 26-position bank levels or one to eight 52-position electrical bank levels	Twelve 11-position levels or six 22-position levels
Coil Power (typical)	25 w	19 w
Dimensions (nominal)	$7 \times 5\frac{1}{4} \times 3\frac{1}{4}$	$4\frac{9}{16} \times 3 \times 3\frac{1}{8}$
Cost: Each, small qty., 12 levels.	$75	$60

†All are indirectly driven with impulse or self-interrupted control, and operate (with proper peripheral equipment) at 65 steps per second. Coil voltages are 6, 12, 24, 48, 60, and 110 vdc; 115 vac with rectifier. Phosphor-bronze contacts standard, gold available. Periodic lubrication required.

9.8 WHO MAKES WHAT

Table 9–9 lists some of the major switch manufacturers. (p. 310)

SUGGESTED READINGS

Dilatush, Earle, "Switches—The 'New' Components Boast a Long Honorable History." *EDN*, April 5, 1975.

Gilder, Jules H., "Focus on Lighted Switches." *Electronic Design, 5,* March 2, 1972.

Golbeck, Bernard J., "Designer's Guide to Rotary Switches," *EDN,* Jan. 5, 1975.

Grossman, Morris, "Focus on Rotary and Thumbwheel Switches." *Electronic Design*, Sept. 27, 1974.

Lockwood, J. P., *Applying Precision Switches, A Practical Guide.* Micro Switch, Division of Honeywell, Freeport, Ill., 1973.

TABLE 9-9 Switch Manufacturers

	Alcoswitch	Arrow-Hart	Automatic Electric	Centralab	Cutler-Hammer	Grayhill	LICON	Mallory	Micro Switch	Oak	Potter & Brumfield	Robertshaw	Shallcross	Stackpole Components	Switchcraft
Snap Switches					x		x		x		x	x			
Rotary Switches				x		x		x		x			x	x	
Lever Switches		x			x										x
Slide Switches	x													x	
Toggle Switches	x	x			x		x		x						
Rocker Switches	x				x									x	
Push-Button Switches															
Plain	x	x			x	x	x		x			x			x
Illuminated					x		x		x						x
Display							x		x						x
Mechanically-Interlocked											x				x
Rotary Stepping Switches				x											

CATALOGS

Automatic Electric (GTE), Northlake, Ill.

Centralab Electronics Division, Globe Union, Inc., Milwaukee, Wisc.

LICON, Division of Illinois Tool Work, Inc., Chicago, Ill.

Micro Switch, Division of Honeywell, Freeport, Ill.

Oak Industries, Inc., Crystal Lake, Ill.

Robertshaw Controls Co., Columbus, O.

Stackpole Components Co., Raleigh, N. C.

Switchcraft, Inc., Chicago, Ill.

CHAPTER TEN

Connectors

Connectors (Fig. 10–1) have been aptly described as sources of trouble joining two pieces of electronic equipment. The cause of the trouble is usually either improperly terminating the connector, or using the wrong connector for the application. Terminating techniques are outside the scope of this handbook. Selecting the right connector requires the consideration of many factors. This chapter contains general information on the families of connectors most widely used in industrial, military, and aerospace electronic equipment.

There is no way all the available connectors could be described in this handbook because there are just too many, nor can the descriptions get too involved in specific contact arrangements and exact dimensions of the lines of connectors (except as examples) without the handbook turning into a catalog.

Most components are identified generically—wirewound resistor, telephone relay, rotary switch, FET, etc.—but many connectors are identified by trade names or military specification.

Families of commercial and industrial connectors are usually developed by a single company to meet a particular need. When additional companies are licensed to manufacture the connector family, new trade names are often devised. In this handbook, these connectors are identified by their original trade names.

Other families of connectors have been developed to meet a military specification, or a military specification is written to cover the connector family. To avoid any possible confusion, these connectors are identified by their Mil Spec.

Fig. 10–1 Connectors. ICBM umbilical cord connector with self-contained explosive charge for blast-off disconnect. Tiny Tim ® micro-miniature connector. *(Courtesy of Bunker Ramo Amphenol)*

Connector symbols are shown in Fig. 10–2. The parts of a typical connector are shown in Fig. 10–3.

10.1 DEFINITIONS

BACK MOUNTED: A connector mounted from the inside of a panel or box with its mounting flanges inside the equipment.

BACK SEAL: The device or material used to provide a hermetic seal over the back end of the contacts of a connector.

BAYONET COUPLING: A quick-coupling device used on circular connectors. Pins on one connector shell engage in spiralled grooves in the mating connector shell; engagement or locking is

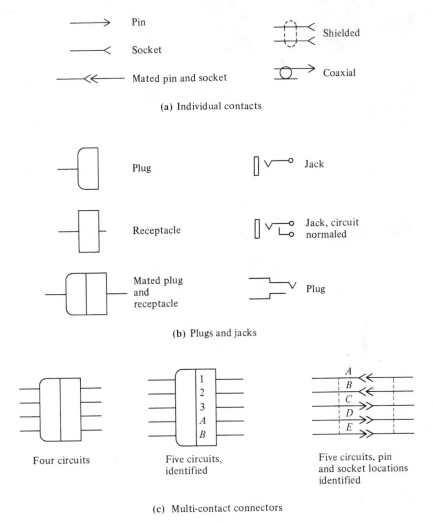

(a) Individual contacts

(b) Plugs and jacks

(c) Multi-contact connectors

Fig. 10-2 Commonly used connector symbols.

accomplished by the partial turn of the bayonet ring while applying axial force against a latching spring.

BLADE CONTACT: A flat, rectangular form of pin contact that mates with a socket contact consisting of opposing spring-loaded jaws that engage both flat sides of the blade.

CABLE CLAMP: Support for conduit, cable or wire at the rear of the connector. It provides strain relief for soldered or crimped connections.

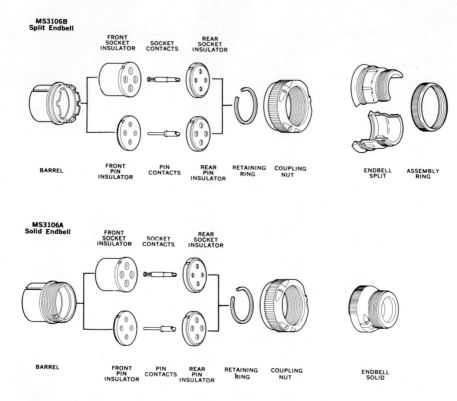

Fig. 10–3 The parts of a typical connector. The MIL-C-5015 connector and its descendents are probably the most widely used electronic equipment connectors in both military and industrial systems. *(Courtesy of ITT Cannon Electric)*

CONFIGURATION: The number, spacing and arrangement of contacts in a multiple-contact connector. See Contact Arrangement.

CONTACT: The mating current-carrying part of a connector.

CONTACT ARRANGEMENT: The configuration of contacts in a connector, including the type of contact (pin, socket, coaxial, etc.), the size or sizes of the contacts, the number of contacts, and the exact physical arrangement of the contacts in the insert. (See POLARIZATION.)

CONTACT RATING: The current-carrying capacity of a mated pair of contacts. For connectors, contact rating is not an interrupt- or make-rating.

CONTACT RETAINER: The device that holds a contact in the insert. It may be a part of the contact or insert. Some retainer designs allow the contacts to be manually removed or inserted in the field by the use of a tool.

CONTACT RETENTION FORCE: The axial load that can be applied to a contact in either direction without its being moved from its normal position in the insert.

CONTACT SEPARATION: In connectors, the physical distance between exposed metal parts of adjacent contacts.

CRIMP: A mechanical method of attaching a wire to a contact by means of indenting or compressing the contact tail against or around the wire to lock or retain the wire within the contact tail.

DIELECTRIC: The insulating material separating the contacts from each other and the shell in a connector. Materials commonly used as dielectrics in connectors are plastic, rubber, glass, and ceramic. Also referred to as insert and insulator.

FACE: The surface of the insert of a connector that faces the equivalent surface of the insert of the mating connector; i.e., the surface of the insert of a pin connector from which the pins protrude.

FRONT MOUNTED: A connector mounted to the outside or mating side of a panel. A front mounted connector can only be installed or removed from the outside of the equipment.

GROMMET: A rubber seal used on the cable side of multiple-contact connectors to seal the connector against moisture, dirt, or air.

HERMAPHRODITE CONNECTOR: A connector that has hermaphrodite contacts or that contains an insert half-pin and half-socket that can be mated with another insert of the same design.

HERMAPHRODITE CONTACT: A contact that is neither pin nor socket and can be mated with another contact of the same design.

HERMETIC SEAL: An air-tight, or more usually, moisture-tight seal between the faces or shells of mating connectors or between a receptacle and the structure to which it is attached. Hermetic seals are of two types: Compressed rubber-like materials, and pin contacts fused in a glass header. Hermetic seals are not perfect, but some are closer to perfect than others.

INSERT: The part of a connector that supports, positions, and insulates the contacts of a connector and mounts them in the shell.

INSERT/EXTRACT TOOL: A tool, usually of frangible plastic, that is used to insert and extract individual contacts from inserts. The tool is designed to break before sufficient force can be exerted to cause any damage to the contact or insert.

INSULATOR: See INSERT.

JACKSCREW: A screw, sometimes with an integral handle, attached to the shell or insert of a rectangular rack and panel connector that is used to draw mating connectors into engagement or to disengage them. Cams are also used for the purpose.

KEY: The projection or projections on the shell of one connector of a mating pair that slides into slots or grooves in the other connector to guide the connectors into proper alignment during mating. (See POLARIZATION.)

PATCH CORD: A short electrical cable with plugs on each end used for rapid and random interconnecting of equipment whose electrical connections are brought out on mating panel-mounted jacks (connectors).

PIN CONTACT: A solid-rod male-type contact designed to mate with a socket or female-type contact. For safety, pin contacts are used in the connector on the load side of the circuit.

PLUG: Any connector attached to a cable or wire. Its contacts may be either pins or sockets.

POKE-HOME CONTACT: A contact, either pin or socket to which a wire has been permanently attached prior to the assembly of the contact into the insert. After the assembly of the wire to the contact, the contact is then "Poked Home" into the insert and retained within the insert by means of retention devices either on the contact or within the insert.

POLARIZATION: The shell configuration or contact arrangement of a connector that guides mating connectors into proper alignment during mating. Polarization also means the mechanical arrangement of keys and slots of connector shells, to bar the mismating of adjacent contactors having the same shell type, size, and contact arrangement. (See KEY.)

POSITION, ALTERNATE POSITION: In circular connectors, the changed angular alignment of inserts in mating connectors from the normal alignment with respect to the shell keys, to provide polarization and prevent mismating of adjacent connectors having the same shell type and size.

REAR INSERTION: A contact and connector design in which prewired contacts can be pushed into the insert from the back of the connector with a special tool. The insert or contact design incorporates a retainer to hold the contact in place; such contacts can also be removed.

RECEPTACLE: A connector permanently mounted on a panel chassis or other structure. Its contacts may be either pins or sockets.

SCOOPING: The damaging of pins of a connector by the inadvertent brushing across them of the shell of the mating connector prior to mating.

Shell: The outer cover of a plug or receptacle that supports the insert and provides mechanical protection for the contacts. The shell, usually made of metal, sometimes also provides the means for coupling and locking mated connectors and provides protection for the lead-wires emerging from the back of the connector. The shell of a panel-mounted connector also provides the means for attachment of the connector to the panel. Also called Body, or Housing.

Socket Contact: A hollow, usually cylindrical, contact that is designed to mate with a pin contact. For safety, socket contacts are used in the mating connector on the source side of the circuit.

Solder Cup: The hollow tubular end of a contact in which a wire is inserted prior to being soldered.

Threaded Coupling: A connector engagement device used on circular connectors containing screw threads. Several complete turns are required to either draw the mating connectors into engagement or lock the connectors together after engagement by straight-line axial force.

10.2 AUDIO CONNECTORS

Audio connectors (Fig. 10–4) are designed for use in audio and other low-level signal applications such as test instruments, computers, and medical instrumentation. The environment for audio connectors is not one of extreme temperature, vibration, and shock, but their environment is rough, with normal hazards of being stepped on, slammed around, tripped over, dropped, and generally mishandled.

Audio connectors can be classed in two grades: Commercial and broadcast. Commercial includes high-fidelity, home tape recording, and amateur use. Broadcast includes use in interconnecting broadcasting, recording studio, television, computer, and medical electronic equipment.

Broadcast Audio Connectors. Cannon XLR/XLRA connectors, and intermateable connectors made by Amphenol and Switchcraft have become the standard microphone and general audio connectors for broadcasting and recording. They are descended from the Cannon XL line introduced in the 1940s. At that time, the standard connectors were the Cannon "P" and "O" types; with the advent of tape recording, television, and more compact transistorized equipment, the smaller and less expensive XL-type connector came into its present wide use.

Fig. 10–4 Audio connectors.

They are made with three, four, or five contacts and in a considerable variety of case styles. Shells are aluminum alloy, die-cast
zinc, or lightweight molded ABS plastic. A thumb-latch provides
quick and positive engagement and disengagement. Shells can float
to or be grounded to one of the contact pins. Contact materials vary.
Some of the connector styles are available with crimp contacts; all
are available with solder pots.

Miniature Audio Connectors. Amphenol 91T, 80, and
Switchcraft Slimline SL connector families are used in commercial,
industrial, and broadcast applications where space is at a premium.
The connectors cannot take the physical abuse that the larger and

stronger XLR/XLRA connectors can withstand. Engagement is by threaded couplings or friction. A great variety of styles and shells is available; contacts range from two to seven. Some connectors in some of the lines match European equipment, but the lines of the different manufacturers are not intermateable.

Phone Plugs and Jacks.   These connectors are used for a lot of things besides plugging in earphones and telephone switchboards. Jack panels and patchcords (short cables terminated at each end with a phone plug) provide unlimited flexibility in interconnecting audio equipment in broadcast and television stations and sound recording studios.

Phone plugs and jacks are made in several sizes, as shown in Fig. 10–5. The 1/4-in. diameter plug is considered to be standard for industrial, military, and telephone audio interconnection. Plugs are also manufactured to a diameter of 0.206 in. to be used in conjunction with 1/4-in. plugs for polarizing (using plugs of different sizes to prevent connection of incorrect equipment). A miniaturized patching system has been developed utilizing a plug diameter of 0.173 in. This system allows twice the number of jacks in a standard 1 3/4-in. by 19-in. rack panel as with the standard-size jacks (Fig. 10–6).

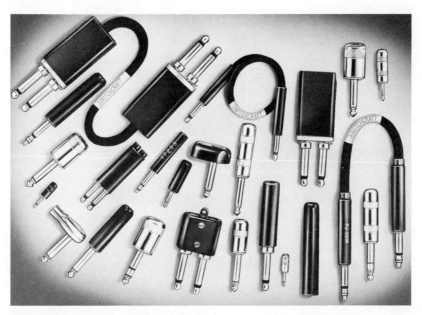

Fig. 10–5 Phone plugs. The connectors are shown one-third size. *(Courtesy of Switchcraft, Inc.)*

Fig. 10–6 Miniature and standard patch panels. Switchcraft series 1600 "tini-telephone" jack panel provides 96 jacks in a standard nineteen inch wide 1 3/4-in. high rack panel compared with 48-in. in a standard series JP jack panel. *(Courtesy of Switchcraft, Inc.)*

For use with miniature and subminiature equipment, two additional plug sizes are made: 0.14-in. and 0.097-in. diameter for miniature and subminiature plugs. These sizes are not generally used for patch panels but rather as individual cable connections.

10.3 RF CONNECTORS

Radio–frequency connectors are made in several series which have dissimilar characteristics. Some of the more common varieties are shown in Fig. 10–7. Most of the time, you don't have much choice in selecting the connector series for a radio-frequency application — in most instances you will use whatever connector family is in the existing interconnected equipment. However, when you do have the choice, there are several factors to consider. For optimum electrical performance (minimum mismatch), the connector size should approximate the cable size. In this way, there are a minimum number of diameter changes in the cable to connector transition. Most con-

Fig. 10–7   Common types of RF connectors. *(Courtesy of Bunker Ramo Amphenol)*

nectors are designed with a nominal impedance of 50Ω and are used with cables of the same impedance. However, these connectors may be terminated with non-50-Ω cables if used at low frequencies (below 150 MHz) where the mismatch due to the connector is negligible.

The voltage and power-handling capability of connectors are dependent on many factors, the most important of which are size and interface and cable entry designs. A higher voltage rating necessitates increasing the creepage path from the center conductor to the outer conductor. This is accomplished by telescoping the interface dielectrics and counterboring the dielectric in the cable entry portion. The latter has a limiting effect on the operational frequency range because of the resulting mismatch in impedance at the connector cable entry. In general, operating frequency range decreases as the connector size increases.

Bayonet connectors have the ability to quickly connect or disconnect a circuit. Threaded connectors generally have better electrical performance and can withstand high vibration levels without producing excessive noise.

Contacts may either be *captivated* or *non-captivated*. The captivated contacts have better mechanical stability, because axial movement caused by temperature changes or extreme cable flexing is prevented. Connectors using non-captivated contacts perform better electrically, are shorter in length, and are cheaper.

Cable connectors terminate two types of cable: Flexible and semi-rigid. Each type of cable has optional termination techniques.

The outer conductor of the flexible cable is terminated either by a clamping device or by crimping. Connectors using the clamping method do not require special assembly tools and connectors may be replaced if damaged without restripping or changing the length of the cable. Crimping requires an assembly tool. The operation requires less time to assemble than a clamp connector and results in a more consistently reliable termination; however, the connector can not be easily replaced without destroying part of the connector or restripping the cable.

The outer conductor of a semirigid cable is terminated by either soldering or clamping. In soldering, the connector body is soldered to the cable jacket with low-temperature solder. The operation is slow and requires skill, but a joint is produced that is excellent both electrically and mechanically. This type of termination is permanent, and the connector cannot be replaced without restripping the cable. Soldered connectors are usually less expensive than clamp connectors, but clamping is a very fast and reliable means of terminating semirigid cable.

The characteristics of common RF connectors are compared in Table 10-1.

BNC Connectors. These connectors are small, weatherproof, lightweight, 50-Ω nominal-impedance bayonet-locking quick-disconnect connectors. The connectors are rated for 500 Vrms at sea level. When terminated with RG55/U, RG58C/U, RG141A/U, RG1428/U, RG223/U, or RG303/U, they offer excellent electrical performance from dc to 4 GHz.

TABLE 10-1   RF Connectors

	Connector Type			
	BNC	TNC	N	UHF
Coupling	Bayonet	Threaded	Threaded	Threaded
Impedance	50	50	50	50
Operating Frequency (max.)	4 GHz	11 GHz	11 GHz	300 MHz
VSWR	1.3 to 1	1.3 to 1	1.3 to 1	—
Working Voltage	500 Vrms	500 Vrms	1000 Vrms	500 Vrms
RG8/U			x	
RG9/U			x	
RG55/U	x	x		
RG58/U	x	x		any
RG87/U			x	coaxial
RG141/U	x	x		cable
RG142/U	x	x		that
RG213/U			x	will
RG214/U			x	physically
RG223/U	x	x		fit
RG225/U			x	
RG303/U	x	x		

Connectors are available with noncaptive and captive center contacts, which are soldered to the center conductor of the cable. The outer shield is terminated by either clamping or crimping.

TNC Connectors.   These connectors are similar to BNC connectors, except that they use a threaded instead of a bayonet coupling. They are small, weatherproof, lightweight, and have a 50-Ω nominal impedance. This feature lets them withstand vibration better and extends the usable frequency range to 11 GHz.

The connectors are primarily designed for use with 50-Ω cables; however, other non-50-Ω impedance cables may be terminated. Both captive and noncaptive contact designs are available. The center conductor of the cable is soldered to the contact, and the outer conductor is secured by means of a clamping mechanism.

N Connectors.   These medium-size 50-Ω connectors are weatherproof, use a threaded coupling, and are rated at 1000 Vrms. An air dielectric is used at the mating face; the center contacts are supported by dielectric beads at the rear of the mating face. They have excellent electrical characteristics from dc to 11 GHz and are considered to be precision connectors. They are extensively used in test equipment.

These connectors are primarily designed for RG8A/U and RG9B/U cables; however, many other cables can be accommodated. Both captive and noncaptive center contacts are available. The center conductor of the cable is soldered to the contact, and the outer conductor is secured by either clamping or crimping.

UHF Connectors. These connectors are low-cost, general-purpose threaded connectors used in low-frequency applications where mismatch is not a problem. They have been extensively used in video applications. They operate satisfactorily up to 300 MHz and have a peak voltage rating of 500 V.

Most UHF connectors have fixed center contacts that are soldered to the center conductor of the cable. Crimp connections are also available.

RF connectors are also made for terminating twin conductor cables used in balanced circuits. A wide range of adaptor connectors are also made to interconnect different types of RF connectors.

10.4 PC BOARD CONNECTORS

Connectors for printed circuit boards are of two types: Receptacles that mate directly with the edge of the PC board or card, and two-piece connector assemblies in which one of the mating connectors is mounted on and hard-wired to the PC board.

PC board edge connectors are available in a wide range of sizes, configurations, contact spacings, and grades (Fig. 10–8). They are used in military, commercial, and industrial applications but not generally in airborne or life-support equipment. Contact can be made with one or both sides of the PC board. The most used contact spacing is 0.156 in., but 0.100, 0.125, 0.150, and 0.200 in. are also standard spacings. Current ratings range from 3 A to 7.5 A per contact. For a given contact spacing, the variations in the number of contacts on a connector are fairly standardized.

Connectors can be polarized by inserting metal or plastic keys into the connector. The keys fit into notches cut in the PC boards.

Several forms of contacts are used in card-edge connectors, as shown in Fig. 10–9. The widely used *bellows* is a folded ribbon that maintains contact well with PC boards of uneven or off-spec. thickness. Wire leads must be soldered. The *ribbon* has characteristics similar to the bellows, costs less to make, and can be terminated by

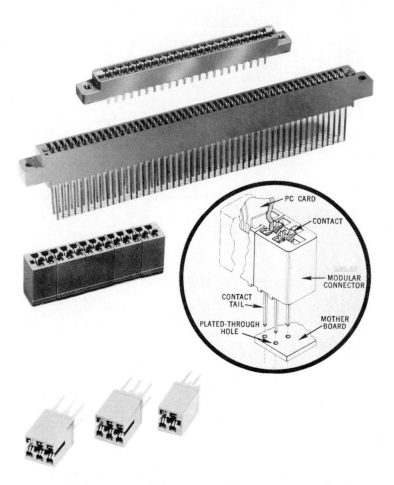

Fig. 10–8 PC board card-edge connectors. (Top) Series 6007 has ribbon contacts, solder tab terminals and 5 A current rating. (Center) Series 6307 has bifurcated bellows contacts with .025 inch square wrapable terminals, 3 A current rating. (Bottom) Series 6308 Mojo® connector is made in four and six contact modules which can be ganged for custom designs. Contact rating is 5 A. *(Courtesy of Elco Corp./Gulf & Western)*

soldering, taper tab, or wire wrap. The *cantilever* contact is less compliant than either the bellows or the ribbon, costs less to make than either, and can be selectively gold-plated or have a gold button welded to the point of contact.

The *tuning* fork contact must contact both sides of the board;

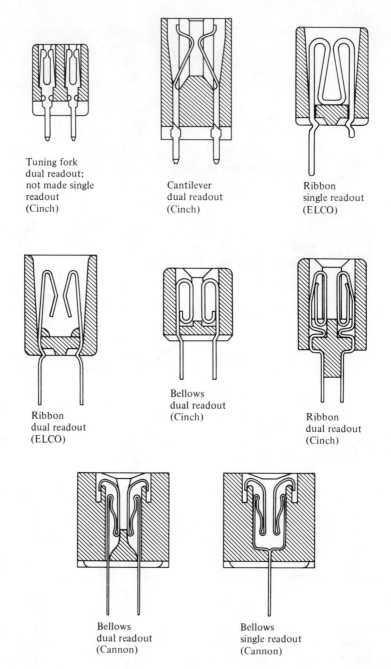

Fig. 10–9 PC board card-edge connector contacts. All are shown enlarged two times.

it cannot handle off-thickness boards, because it has little compliance. It is the cheapest to make. All of the contact forms are available in bifurcated versions.

Two-piece PC board connector assemblies come in two styles: Enclosed connectors and open contacts. Enclosed connectors cost more than either card-edge or open contacts but have longer life, higher contact density, better reliability than card-edge, and lower engagement and withdrawal force than either. Some connectors are made for this application, using tuning fork and blade contacts. Others, like the D-subminiature type, are adapted to the use.

The best known open contact type is Elco's Varicon ® (Fig. 10–10) which is the basic style of hermaphrodite contact used in all of their connectors. These forked contacts mate along four beveled surfaces that are coined for hardness and smoothness. Individual contacts are attached to PC boards by staking or soldering. Various mounting configurations allow PC boards to be mounted on other PC boards. There is considerable flexibility in layout, but contact density is limited.

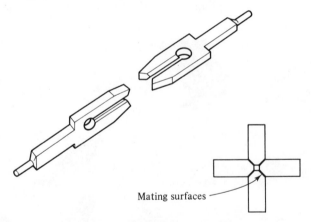

Mating surfaces

Fig. 10-10 Elco Varicon ® hermaphrodite contact. Contacts are used in multi-contact connectors and individually staked to PC boards. *(Courtesy of Elco Corp./Gulf & Western)*

10.5 MINIATURE AND SUBMINIATURE RACK AND PANEL CONNECTORS

These connectors are designed to save space and weight. They are especially suited for aircraft, missile, satellite, and ground support systems. Although designed for rack/panel applications, they are also widely used as cable connectors with accessory junction shells (shields) with integral clamps.

D-Subminiature Connectors.   These connectors (Fig. 10–11) are made by several manufacturers under different trade names. All are intermateable.

> Cannon–ITT (D-Subminiature)
> Cinch–TRW (D-Subminiature)
> Amphenol (Min–Rac)
> Matrix Science (MD Subminiature)

The connectors are made in four shell sizes with contact arrangements having up to 50 contacts rated at 5 A per contact for #20 wire, with normal spacing. High-density arrangements with 78 size-22 contacts and double-density arrangements with up to 100 contacts are also made. Other arrangements also include mixtures, coaxial, power, and high-voltage contacts. Polarization is by the keystone shape of the shell; coupling is by friction. Locking accessories are available.

Fig. 10–11   D-Subminiature connectors. *(Courtesy of ITT Cannon Electric)*

Original D. These connectors utilize a two-piece nylon insulator in five shell sizes containing 9, 15, 25, 37, and 50 size-20 solder pot contacts that will accommodate wire sizes up to #20 AWG stranded.

Burgun-D Mark IV. These nonenvironmental connectors have crimped rear release snap-in contacts. They are used primarily in commercial applications. The crimp contacts can accommodate wire sizes from #20 to #30 AWG stranded. An expendable plastic tool is used for contact insertion and removal.

D*M Golden-D Mark I (Mil–C–24308). These connectors have solder pot contacts in a monoblock insulator, giving improved temperature performance and better moisture resistance. Insulation is green diallyl phthalate. Contact arrangements include the five standard arrangements, plus 17 additional combination arrangements with unfilled cavities for power, high voltage, or coaxial contacts. These contacts must be ordered separately and may be used interchangeably in any of the combination arrangements.

D*MA Royal-D Mark III (Mil–C–24308). These connectors use a contact assembly that permits rear insertion and release and extraction of contacts.

Double-Density D-subminiature. These connectors use the same shells as D-subminiature connectors but have twice as many contacts in each shell size, using twist pin contacts. In this design, contact positions are reversed, flexible pin contacts are recessed, and the more rugged sockets are exposed. Contacts will accommodate wire sizes from #22 to #26 AWG stranded; the current rating is 5 A per contact. Contact arrangements for these connectors are shown in Fig. 10–12.

Blue Ribbon® Connectors (Amphenol). These rack and panel plugs and receptacles are available with 8, 16, 24, and 32 solder tab contacts rated at 5 A. The ribbon contacts provide smooth, easy insertion and extraction with a self-wiping action that enhances reliability. Barriers provide polarization. Insulation is diallyl phthalate. Plug contacts are gold-flashed copper; receptacle contacts are gold-flashed phosphor bronze. Mounting plates are stainless steel. The shells provide polarization.

Micro Ribbon® Connectors (Amphenol). These connectors (Fig. 10–13) are used both for rack/panel connections and as cable

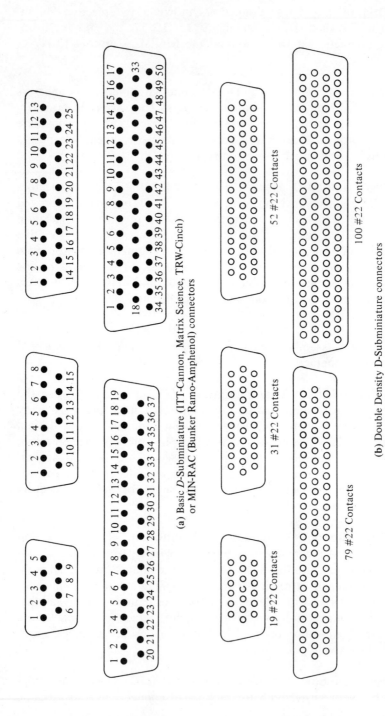

(a) Basic D-Subminiature (ITT-Cannon, Matrix Science, TRW-Cinch) or MIN-RAC (Bunker Ramo-Amphenol) connectors

(b) Double Density D-Subminiature connectors

Fig. 10–12 D-Subminiature connectors. Typical contact arrangements.

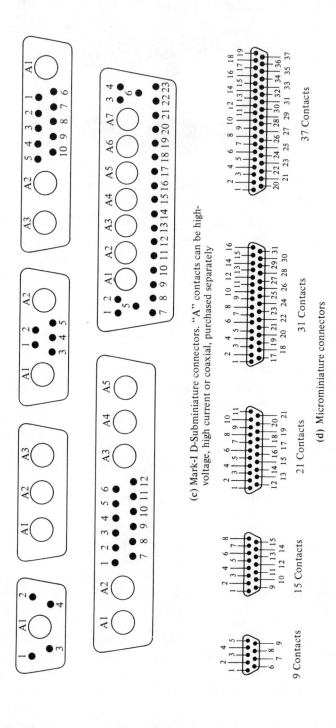

(c) Mark-I D-Subminiature connectors. "A" contacts can be high-voltage, high current or coaxial, purchased separately

(d) Microminiature connectors

Fig. 10–12 *continued*

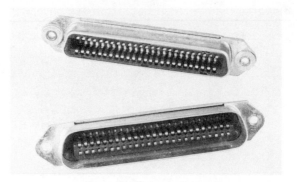

Fig. 10-13 Micro Ribbon connectors.

connectors, accessory hoods, and cable clamps. Contacts are gold-plated cadmium bronze, insulators diallyl phthalate, shells zinc-plated brass. Connectors of various lengths provide 14, 24, 36, and 50 solder tab contacts rated at 5 A per contact. The shell provides polarization.

10.6 STANDARD RACK AND PANEL CONNECTORS

These connectors (Fig. 10-14) are used to interconnect rack/panel chassis-mounted equipment. One-half of the connector assembly is mounted on the stationary equipment rack or panel, and the other half of the assembly is mounted on the removable piece of equipment, allowing easy removal of the equipment for inspection, repair, or interchange. Rectangular connectors are available with a variety of contact arrangements, with individual contact current ratings from 2 A to 101 A. There is practically no intermating between the connectors of different manufacturers. Some are available with shrouds converting the basic chassis connector to a cable plug.

All of these connectors require some device or design feature to assist in mating the connectors. Otherwise, the engaging force would be too much for unassisted hand-mating.

Some connectors have cams, levers, threaded jack-screws, or trick contacts that mate after engagement. To engage some other connectors, you will have to arrange some means to lever your chassis, which is often preferred to a back of the rack or otherwise inaccessible integral engaging device. Table 10-2 compares the characteristics of a diverse group of rectangular rack/panel connectors. Figure 10-15 shows some of the contact arrangements available for a typical connector.

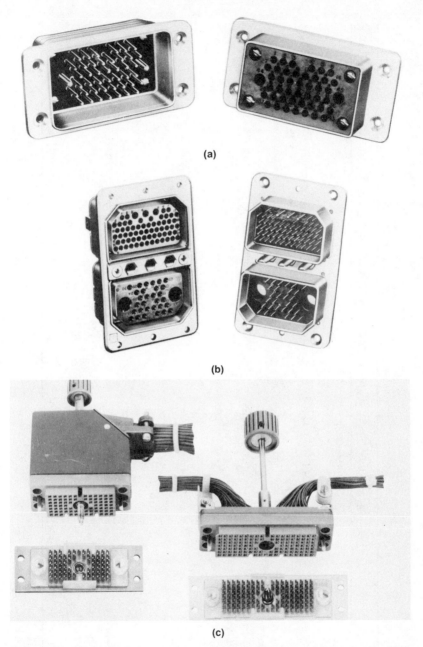

(a)

(b)

(c)

Fig. 10–14 Standard size rack and panel connectors. **(a)** Cannon DPD, **(b)** Cannon DPX. Both are available with a wide choice of contact ratings and types. **(c)** Elco 8026, available with Varicon® only. *(Courtesy of ITT Cannon Electric; Elco Corp./Gulf & Western)*

TABLE 10-2 Characteristics of Typical Rectangular Rack and Panel Connectors

	Cannon DPK	Cannon DPX	Cannon DPD/DPDMA
Application	Military Mil-C-83733 (USAF) −54°C to +200°C	Industrial −67°F to +257°F	Industrial −55°C to +125°C
Contacts	Rear insert and release	Rear insert and release	Rear insert and release
Termination	Crimp	Crimp or solder	Crimp or solder pot
Contact Material	Gold-plated copper alloy	Gold-plated copper alloy	Gold-plated copper alloy
Number of Pin Contacts	18 to 185	2 to 106†	10 to 128†
Sizes and/or Current Rating	12, 16, 20, 22	12, 16, 20	4, 6, 8, 10, 14, 16, 20
Other Contacts	Coaxial	Coaxial	—
Shells	Diecast aluminum	Aluminum alloy	Aluminum alloy
Finish	Nickel	Cadmium plate	Cadmium plate
Insulators	Glass-filled epoxy	Melamine, phenolic, epoxy, diallyl phthalate	Melamine, phenolic, diallyl phthalate
Seals	Silicone rubber	None	None
Mating Assist	Jackscrew	None	Jackscrew in two-gang versions
Plug Covers	No	Yes	Yes
Polarizing Positions	6	99 (Two-gang only)	36

† Double the quantity in two-gang versions.

TABLE 10-2 *Continued*

	Cannon DL	*Elco 8026*	*Cannon DJS*	*Cannon DPGM/DPJM*
Application	Industrial, low-cost −65°C to +100°C	Industrial, military	Miniature Mil–C–38999 −65°F to +302°F	Missile, aircraft, ground support
Contacts	Hermaphrodite	Hermaphrodite Varilock®	Removable	Rear removable
Termination	Crimp or wirewrap	Solder, taper tab, crimp, or wirewound	Crimp	Crimp
Contact Material	Gold-plated copper alloy	Gold/nickel-plated phosphor bronze	Gold-plated copper alloy	Gold-plated copper alloy
Number of Pin Contacts	60 to 156	33, 55, 75, 79, 117	78 or 128	8 to 98
Sizes and/or Current Rating	5A	3 A	22 3 A	4, 8, 12, 16, 20 75 to 80 A
Other Contacts	—	—	—	Coaxial
Shells	Polyvinylchloride	None	Stainless steel	Aluminum alloy
Finish	—	—	—	—
Insulators	Glass-filled noryl	Diallyl phthalate, polycarbonate, nylon	Glass-filled epoxy	Diallyl phthalate
Seals	None	None	Silicone	Polychloroprene
Mating Assist	Zero engaging force contacts mate after engagement	Jackscrew	Jackscrew	None
Plug Covers	Yes	Yes	No	No
Polarizing Positions	36	36	6	1

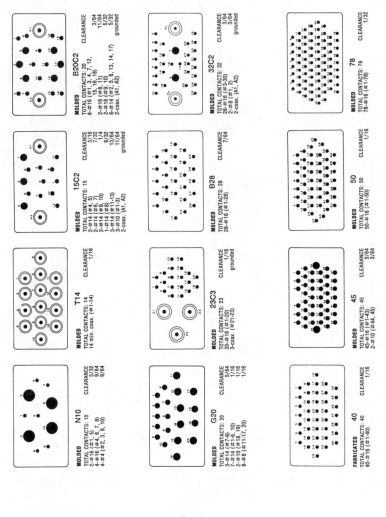

Fig. 10-15 Cannon DPD rectangular rack and panel, typical contact arrangements. Shown one-third size. *(Courtesy of ITT Cannon Electric)*

10.7 STANDARD THREADED CIRCULAR CONNECTORS

This family of connectors began as "AN" connectors (Army–Navy) and were originally designed for aircraft use. They are now widely used in military, aerospace, and industrial applications because of their low cost; uniform high quality; broad range of sizes, styles, variations, and pin arrangements; and ready availability from many vendors, including mail-order distributors. Figure 10–16 shows two of the many varieties.

Fig. 10–16 Standard threaded circular connector, MIL-C-5015. *(Courtesy of ITT Cannon Electric)*

Specialized types have evolved from the basic design to meet specific new requirements. All types are interconnectable, intermountable, and interchangeable. (Some manufacturers make proprietary contact arrangements that are not MS approved and that would not work with contact arrangements of other vendors.)

Shell Size and Contact Arrangement. Shell sizes are based on the diameter of the coupling threads in sixteenths of an inch. For example, a size-18 shell has a diameter of 1-1/8 in. (18/16 in.); a size-24 shell has a diameter of 1-1/2 in. The type and number of contacts in a connector are defined by a combination of the shell size and a contact arrangement number, for example 18–22, 32–8, etc. The right-hand number is the contact arrangement number and tells you nothing about the number of pins or their size. The 18–22 example happens to have three #16 pins, and the 32–8 has twenty-four #16 plus six #12 pins. You must consult the chart in the catalog.

Inserts are available to provide polarization in order to prevent cross-plugging adjacent, otherwise identical, connectors. The inserts are rotated to different standard positions within the shell relative to the shell key.

Contacts. Although they will intermate, the contacts in the different classes of connectors are manufactured in different ways, and they have different current ratings (Table 10–3). In some connector classes, you can obtain contacts machined from alloys to match thermocouple lead-wires.

TABLE 10-3   Contact Current – Carrying Ratings for
Miniature Circular Connectors

	Current, Amperes		
Contact Size	Standard & Environmental Connectors	Firewall Connectors	Hermetically Sealed (Glass-Header) Connectors
18	22	22	10
12	41	41	17
8	73	73	33
4	135	135	60
0	245	245	100

Mil–C–5015. Contacts are silver-plated copper alloy and have tinned solder pot terminals. Standard insert insulators are molded from a rigid plastic such as diallyl phthalate. For extreme environments and high altitude, resilient insulators and wire-sealing grommets are available that will provide protection from condensation, vibration, corona, and flashover.

Shells are cadmium-plated aluminum, finished in olive drab. For resistance to salt water, fuels, etc., and vibration, some connectors can be obtained with plastic potting cups. Accessories include junction shells, protective caps, dummy receptacles, cable clamps, and telescoping bushings.

Fireproof (Mil–C–5015). Fireproof, or firewall connectors provide a means for penetrating the engine firewall of military and commercial aircraft with electrical circuits while still maintaining the flame-barrier integrity of the firewall. These connectors provide protection against normal high operating temperatures, emergency fire-retardant conditions, moisture, and atmospheric changes and are resistant to fuels, cleaning agents, coolants, and hydraulic fluids. Construction is shown in Fig. 10–17.

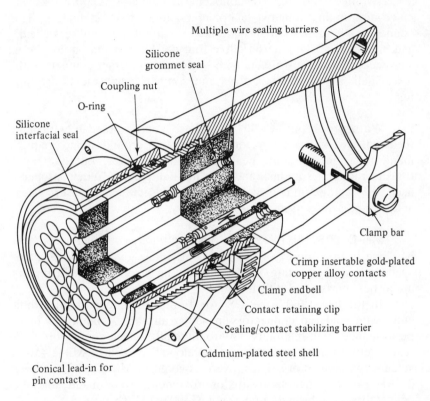

Fig. 10–17 Fireproof MIL-C-5015 connector construction. *(Courtesy of ITT Cannon Electric)*

In test, a mated connector must carry the maximum rated dc current for five minutes without a break in continuity and must prevent the passage of flame for 20 minutes while the connector is subjected to a +1093°C flame and is being vibrated at 33 Hz. During the sixth minute, with 115 vac 60 Hz applied, leakage between contacts and shell and between adjacent contacts won't exceed 2 A.

Shells are steel, finished in olive drab. A typical insulation material is glass-filled melamine. Grommets, if used, are silicone. Contacts are crimped to lead-wires. Shell size and contact arrangement are limited.

Hermetically Sealed (Mil–C–5015). Hermetically-sealed receptacles are used where a partial vacuum, inert gas, or constant controlled pressure are used inside a piece of equipment to eliminate adverse effects created by atmospheric changes. Applications include aircraft instruments, tachometers, and direction finders. The connector's compression-glass hermetic seal will prevent an air leakage in excess of one micron cubic foot per hour at one atmosphere. The receptacles are available with solder pot or eyelet contact terminals. Shells are steel. Shell sizes and contact arrangements are limited.

Waterproof (Mil–C–5015). These connectors can withstand the mud, ice, and water encountered around ground support equipment, radar installations, heavy construction sites, outdoor rapid transit, radio/tv stations, and marine equipment. When properly used, they can be immersed. An O-ring seals the mated insert faces, and a gland seal is used at the cable entry.

Mil–C–005015F (Navy). Soldering wire to connector contacts has long been a source of trouble; cold solder joints, wicking, damaged insulation, shorting, etc., reconnecting errors, and field repairs are at best messy and difficult.

In the late 1960s, the US Air Force ran into particular difficulty with soldered connectors in newer aircraft. A family of high-performance, high-reliability, environmentally sealed connectors with rear-inserted crimp contacts was developed to replace Mil–C–5015 solder-type connectors. These were covered in Mil–C–83723 Series 2 (Air Force). Shell hardware improvements were made, and the connectors were brought under Mil–C–005015F (Navy).

This specification incorporates the rear release contact retention system and fail-safe plastic insertion/extraction tools. If the tool is used improperly, it will break before damaging the connector (Fig. 10–18).

These contacts ease maintenance, allow use of common terminating tools, and minimize improper field repair. They are made in four classes: General-purpose, high-temperature (200°C), high-temperature fluid-resistant, and firewall.

Sockets have hard-plastic faces with lead-in chamfers that will not accept a pin contact that is bent beyond pre-established limits. The silicone interface seal on the pin connector has raised barriers around each pin that displace into the mating chamfer when the connectors are mated to provide a moisture seal. Silicone grommets provide protection from fuel, coolants, cleaners, and hydraulic fluids.

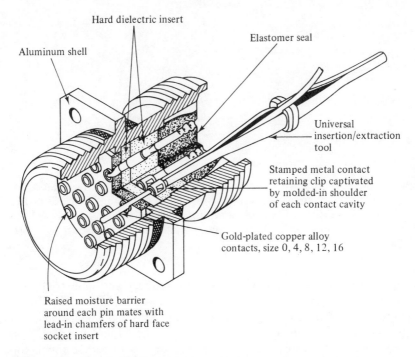

Hard dielectric insert

Elastomer seal

Aluminum shell

Universal insertion/extraction tool

Stamped metal contact retaining clip captivated by molded-in shoulder of each contact cavity

Gold-plated copper alloy contacts, size 0, 4, 8, 12, 16

Raised moisture barrier around each pin mates with lead-in chamfers of hard face socket insert

Fig. 10–18   MIL-C-00515F (Navy) standard threaded circular connector construction. *(Courtesy of ITT Cannon Electric)*

Connectors to Mil–C–00501F include aluminum-shelled plugs and receptacles in all standard versions, firewall connectors, and a push-on/pull-off quick-disconnect plug.

10.8 MINIATURE BAYONET CIRCULAR CONNECTORS

Bayonet connectors (Fig. 10–19) can be mated and demated faster than equivalent connectors with threaded couplings. Not all varieties are intermateable.

Mil–C–26482. These versatile quick-connect connectors are hermetically sealed and have a three-point bayonet coupling and five-key polarization. The miniature circular connectors are grouped into three series: General-purpose with solder contacts for military and commercial applications, general-purpose with crimp contacts for industrial applications, and high performance environment-resis-

Fig. 10–19 Miniature bayonet circular connectors can be mated and demated faster than equivalent threaded connectors. MIL-C-83723B connector shown is high temperature, fluid resistant type, has 55 #20 contacts. Connector is shown approximately full size. *(Courtesy of ITT Cannon Electric)*

tant connectors with crimp contacts for military and industrial applications (Fig. 10–20). All have cadmium-plated aluminum-alloy shells with olive drab chromate finish, polychloroprene resilient insulators, grommets, and seals. Crimp contacts are inserted and removed from the rear, but the tools used are not the same as for other connectors described in this section and in Section 10.7.

Connectors for special applications are also available. They include RFI filtering versions with low-pass internal filter pin contacts and glass header hermetic connectors for high pressure watertight requirements.

All types are intermateable. Termination hardware is physically interchangeable but not with all styles.

Mil–C–83723 Series 1. These connectors intermate with the Mil–C–26482 style. Pin and socket crimped contacts are inserted and removed from the rear of the connector using a fail-safe plastic tool. Both square-flange and single-hold mounting-type receptacles and the mating plugs are available in a wide range of shell sizes and insert arrangements.

Mil–C–83723 Series 3. This family of connectors was developed to provide the aerospace and aircraft industries with the highest possible reliability and environmental sealing in connectors. The design is foolproof for both experienced and inexperienced operators during installation, modification, and field repair. These connectors all have rear-release contacts, inserted and removed with a fail-safe plastic tool.

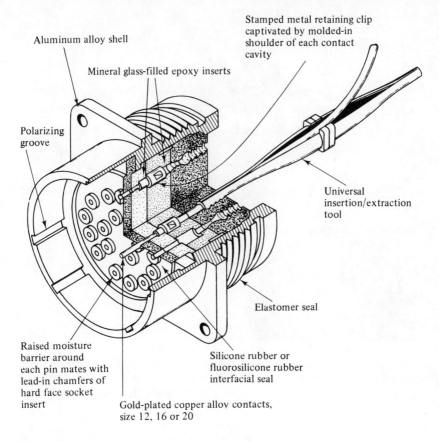

Fig. 10–20 Miniature bayonet circular connector construction. *(Courtesy of ITT Cannon Electric)*

Sockets have hard-plastic faces with lead-in chamfers that will not accept a pin contact that is bent beyond pre-established limits. The silicone interface seal on the pin connector has raised barriers around each pin that displace into the mating chamfer when the connectors are mated to provide a moisture seal. Silicone grommets provide protection from fuel, coolants, cleaners, and hydraulic fluids.

10.9 HIGH-DENSITY QUICK-DISCONNECT CIRCULAR CONNECTORS

Designated ASTRO–348 ® by Amphenol, this family of sub-miniature connectors (Fig. 10–21) provides greater contact density

Fig. 10–21 High-density quick-disconnect circular connector. Amphenol ASTRO-348 line can provide 155 #23–22 contacts in a size 24 shell. Connector shown is one-half size. *(Courtesy of Bunker Ramo Amphenol)*

than is available in miniature circular connectors. This is made possible by an insert design that replaces metal contact retention clips with a one-piece molded insulator contact retention system and Poke-home® contacts that save space. An additional space saving feature is the use of smaller pins.

These connectors are made in two series using different shell designs: Series 1 (long) and Series 2 (short). Although both series use the same inserts, the two series cannot be intermated, because they have different shell key positions.

Series 1 connectors are "no-scoop," which means it is impossible to damage pin contacts in the connector when the mating connector shell is accidentally wiped across its face, as happens in blind-mating connectors, even in an attempt to make a pin plug to a pin receptacle. Series 1 connectors also have the interior ringed with shell-to-shell grounding fingers for improved EMI protection. If the shorter standard Series 2 connectors are used, as is normal, with the pins in the receptacle, no-scoop protection is also obtained.

10.10 MICROMINIATURE CONNECTORS

Miniaturization and increasing complexity of aircraft, space vehicles, and electronic data processing equipment have required an equivalent reduction in connector size. As a result, microminiature connectors (Fig. 10–22) are amongst us, with healthy price tags compared to equivalent subminiatures.

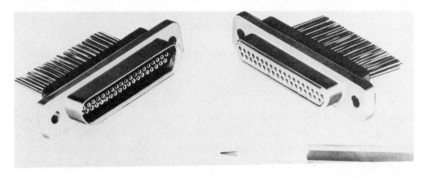

(a)

(b)

Fig. 10–22 Microminiature connectors. **(a)** Amphenol 222 series Mighty-Mite® has 7 to 61 contacts on 0.085 centers *(Courtesy of Bunker Ramo Amphenol)*. **(b)** Cannon MDM series offers 9 to 51 contacts on 0.050 centers *(Courtesy of ITT Cannon Electric)*.

Typical of these connectors, the Cannon MDM connectors have contacts on 0.05-in. centers instead of 0.1-in. and use a unique Micropin/Microsocket® contact combination to achieve the close spacing. In this contact design, the pin is the spring member, mating with a rigid socket (Fig. 10–23). The pin consists of seven gold-plated copper alloy spring wires twisted into a miniature cable. The

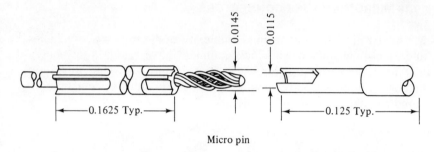

Micro pin

Fig. 10–23 Micropin and Microsocket®

cable is crimped in a terminal at the back end and fused together at the exposed tip in a weld. The wire bunch is compressed axially causing the individual spring wires to bulge outward. The expanded cable now provides seven spring members peripherally around the pin so that no matter what radial forces are applied, contact is maintained with the mating socket wall.

The flexible twist pin is recessed into a cavity in the insulator, and the rigid socket is exposed, reversing the traditional positions of pin and socket. During mating, the socket is guided into the pin cavity by the lead-in chamfer and surrounds the pin. The pin is kept from flexing beyond the socket capture radius by the walls of the cavity. The hemispherical weld with which the pin is capped combines with the lead-in chamfers on the sockets and socket insulator to guide the pin into alignment for positive mating.

Microminiature Rectangular. (Cannon MD/MDM Series, Cinch Dura-Con D Series) These connectors resemble D-subminiature connectors and have the same contact arrangements—9, 15, 25, 37, and 51 contacts—plus a 21 and 31 contact size. A version with coaxial cable contacts is also made. Contacts are terminated in flexible #25 AWG solid conductor uninsulated pigtails for direct soldering into circuits, or with 18-in. #26 Teflon insulated stranded wires.

Integral insulation and shells are glass-filled diallyl phthalate, nylon, or polycarbonate. The connectors are also made in a version with cast aluminum-alloy shells. To overcome the half-ounce per contact mating force, jackscrews are used to mate connectors.

Microminiature Round Connectors. (Cannon Micro–K) These threaded connectors are made in two shell sizes with only one contact arrangement for each. Connectors are supplied with 18-in. #26 AWG insulated wires attached to the contacts.

10.11 UTILITY RECTANGULAR CONNECTORS

These connectors are used both as rack and panel connectors and as cable connectors.

Jones Connectors. (Cinch-TRW) For more than a generation, "Jones" (no relation) plugs and sockets have been the standard low-cost multicontact connector in the electronic/electrical industry. There is no area in which they haven't been used—amateur radio, consumer electronics, broadcasting, industrial control and automation, test instruments, vending machines, and aerospace. The connectors are made in two sizes. Insulating material is black phenolic. Phosphor-bronze double-wipe knife-switch contacts make contact with both sides of the brass plug blade contacts.

Varicon Connectors. (ELCO) These are do-it-yourself connectors. Each connector is assembled from three basic parts: Hermaphrodite contacts, insulator end sections, and insulator center sections. Parts are purchased in bulk. They are held together with through-bolts and are mounted and enclosed with a selection of brackets and covers. Parts can be quickly assembled to make connectors equivalent to Jones connectors but with almost unlimited variation and polarization.

10.12 HYDROSPACE CONNECTORS

Hydrospace connectors (Fig. 10–24) are designed to meet both military and commercial requirements for connectors able to withstand the environments encountered by submarines and other undersea vehicles. These include pressures up to 10,000 psi and long periods of saltwater immersion.

Brass Shell Mil–C–5015. These connectors are rated to 4500 psi (10,000 feet or 1667 fathoms). They use heavy brass shells and a limited range of standard Mil–C–5015 contact arrangements. Watertight coupling between mated connectors is obtained with Acme threaded coupling and rubber sealing ring. Cable clamps and Kellums grips provide cable strain relief.

Fig. 10-24 Hydrospace connectors. These MIL-C-24217 connectors are capable of withstanding pressures up to 10,000 psi. *(Courtesy of ITT Cannon Electric)*

Mil-C-24217. These hydrospace connectors are designed to withstand pressures up to 10,000 psi. Shells are stainless steel with aluminum-bronze coupling nuts. Plugs have diallyl phthalate insulators and gold-plated copper alloy contacts. Size-16 contacts are crimped; sizes-12, 8, and 4 are solder pot and are sealed with individual O-rings. Hermetically-sealed receptacles have compression glass insulators and gold-plated mild steel contacts. Receptacle shells come in several styles depending on mounting requirements: Weld mount, bolt flanged, lock nut, feed-thru union, or in-line, as shown in Fig. 10-25.

Mil-C-22249. These connectors feature bulkhead receptacles with feed-thru pin-to-pin contacts mated on the internal side of the bulkhead with a low-pressure plug and on the outside with a high-pressure plug. The feed-thru receptacles can withstand 10,000 psi unmated. The plugs in most sizes have rear-inserted contacts. As many as 209 contacts can be obtained in one connector. Plug contacts are gold-plated copper alloy; receptacle contacts are molybdenum or nickel-clad molybdenum. Shells are stainless steel with a manganese-bronze coupling nut.

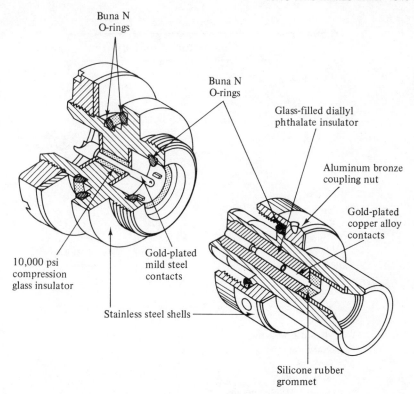

Fig. 10–25 Construction of MIL-C-24217 connectors. *(Courtesy of ITT Cannon Electric)*

10.13 WHO MAKES WHAT

Table 10-4 lists some of the major connector manufacturers. (p. 350)

CATALOGS

Bunker Ramo Amphenol, Broadview, Ill.

Elco Corporation (Gulf and Western), El Segundo, Calif.

ITT Cannon Electric, Santa Ana, Calif.

Matrix Science Corp., Torrance, Calif.

Switchcraft, Inc., Chicago, Ill.

TRW Cinch, Elk Grove, Ill.

TABLE 10-4   Connector Manufacturers

	Amphenol	ITT Cannon	TRW Cinch	ELCO (Gulf & Western)	Matrix Science	Switchcraft
Audio Connectors						
Broadcast	x	x				x
Miniature	x	x				x
Phone Plugs, Jacks						x
RF Connectors						
BNC	x	x				
TNC	x	x				
N	x	x				
UHF	x	x				
PC Board Connectors						
Board Edge	x	x	x	x		
Two-Piece	x		x			
Open Contact				x		
Miniature & Subminiature						
Rack and Panel Connectors						
D-Subminiature/Minrac-Style	x	x	x		x	
Blue-Ribbon/Micro-Ribbon	x		x			
Standard Rack and Panel						
Connectors, All Kinds	x	x		x		
Miniature Threaded Circular						
Connectors, All Kinds	x	x			x	
Miniature Bayonet Circular						
Connectors, All Kinds	x	x	x		x	
Standard-Sized Circular						
Connectors, All Kinds	x	x				
High-Density Quick-Disconnect						
Circular Connectors,						
All Kinds	x	x	x			
Microminiature Connectors,						
All Kinds	x	x				
Utility Rectangular						
Connectors, All Kinds			x	x		
Hydrospace Connectors,						
All Kinds		x				

CHAPTER ELEVEN

Semiconductors

Any electronic component whose main functioning parts are made from a semiconductor material is called a *semiconductor*. It is also called a solid-state device, because it has no moving parts, as does a relay, or a switch, or a potentiometer.

Semiconductors (Fig. 11–1) are *active* components in that they are capable of controlling currents or voltages to produce gain or switching action in a circuit. Relays and vacuum tubes are also active components. Resistors, capacitors, and transformers are examples of *passive* components.

New semiconductor products are introduced almost daily, and new applications for semiconductors increase at an even faster rate. It would be impossible to include descriptions of all varieties in one chapter of one handbook; this chapter contains descriptions of only the major classes of one-function discrete semiconductor components, transistors, diodes, and thyristors. Voltage variable capacitors (diodes) are covered in Chapter 2. Typical power transistor case styles are shown in Fig. 11–2.

Theory. Semiconductors are materials that lie somewhere between metals and insulators in their ability to conduct electricity. Germanium and silicon are two such semiconductors. The atoms of both materials have four electrons in the outer shell. Atoms of germanium and silicon form crystals in which each atom shares these four valence electrons with four neighboring atoms in covalent bonds. As there are no free electrons to move in the crystal, the crystal is a poor electrical conductor. Conductivity can be increased

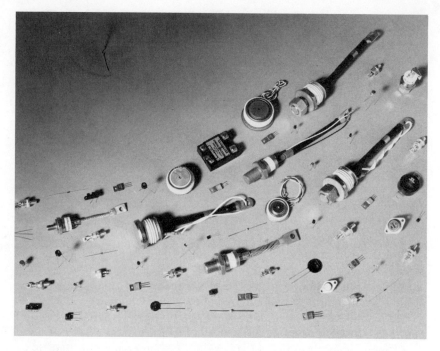

Fig. 11-1 Semiconductors. *(Courtesy of General Electric Co.)*

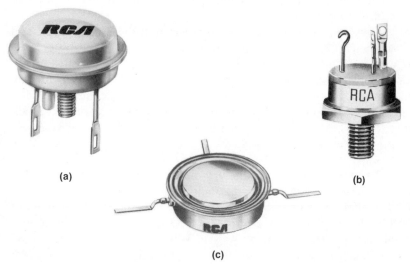

(a)

(b)

(c)

Fig. 11-2 Typical power transistor case styles. **(a)** TO-36, **(b)** TO-63, **(c)** TO-220AB. *(Courtesy of RCA.)*

by heating the crystal, which vibrates the atoms causing one of the valence electrons to occasionally break away and wander through the crystal. The electron-deficient atom now has a positive charge equal to the charge of the missing electron and can now attract an electron from a neighboring atom, which can then acquire one, and so on. Each free electron breaking away will produce an electron deficiency that can move through the crystal as readily as the free electron itself. These deficiencies can be thought of as particles with positive charges and are called *holes*.

The conductivity of a semiconductor can also be increased by adding impurities to the semiconductor crystal when it is formed. The process is called *doping*. Atoms of these impurities replace germanium or silicon atoms in the crystal structure. These impurities may be either donors or acceptors. Phosphorus, arsenic, and antimony are donor materials; their atoms have five valence electrons, four of which form covalent bonds with neighboring semiconductor atoms, leaving one free electron that can be easily freed from the atom and can then move through the crystal. The donor atom assumes a positive charge but cannot move in the crystal. A semiconductor containing donor atoms is an N-type semiconductor, because conduction is by means of free electrons (negative charge).

Boron, aluminum, gallium, and indium are acceptor materials; their atoms have three valence electrons, which all form covalent bonds with neighboring semiconductor atoms. The missing fourth electron leaves a hole in the crystal. The acceptor atom has a negative charge and can acquire an electron, and the hole can move through the crystal. A semiconductor containing acceptor atoms is a P-type semiconductor, because conduction is by means of free holes in the crystal (positive charge).

In summary, conduction in semiconductor materials takes place by means of free electrons and free holes. The electrons and holes originate from thermal generation of hole–electron pairs or from donor or acceptor impurities. The conductivity of a semiconductor crystal can be controlled during manufacture. An N-type or P-type can be produced by adding controlled amounts of donor or acceptor impurities.

If a P-type region and an N-type region are formed in the same crystal structure, a rectifier or diode is produced. The boundary between the two regions is called a *junction*.

If the P-region is made positive with respect to the N-region by an external circuit, then PN junction is *forward-biased,* and the junction has a very low resistance to the flow of current. Holes in the positive P-type material are attracted across the junction to the negative side, and the free electrons in the N-type material are likewise attracted to the positive side.

If a positive voltage is applied to an N-zone with respect to the P-zone terminal which is made negative by an external circuit, the PN junction is *reverse-biased*. At voltages below the breakdown voltage of the junction, the current flow may be cut off, or only a small leakage current flows, which is called the *reverse saturation current*. The positive holes and the free electrons are repelled from the junction by the polarity of the applied voltage, leaving the region adjacent to each side of the junction free of charge carriers. Such regions are called *depletion regions*. But, the presence of acceptor ions fixed in the lattice on the P-side will give the depletion region on that side negative charge; donor ions fixed in the N-side give that side a positive charge. These opposing regions of charged ions create a strong electric field across the PN junction. This electric field generates the reverse current.

In addition, because semiconductor regions are never perfect, all P-type material contains some free electrons, and all N-type material contains some holes. Thermally generated electron hole pairs also contribute to the reverse saturation current which, at a fixed temperature, is relatively independent of voltage.

The PN junction, if not reverse-biased beyond the breakdown voltage, conducts heavily in one direction only and functions as a rectifier. Increasing the reverse-bias voltage to the breakdown voltage results in a rapid increase in current flow. The change from a low current to high current is abrupt (Fig. 11–3); the transition is called the *zener knee*. With increasing reverse voltage beyond breakdown, the voltage drop across the PN junction remains essentially constant for a wide range of currents; this region beyond breakdown is the *zener control region*.

There are two breakdown mechanisms: zener and avalanche. In *zener* breakdown, the value of the breakdown voltage decreases as junction temperature increases. In *avalanche* breakdown, the value of the breakdown voltage increases as the junction temperature increases. The relative concentrations of impurities in the two junction materials determines which form of breakdown mechanism occurs.

A junction with a wide depletion region will breakdown by avalanche; a junction with a narrow depletion region will breakdown by the zener mechanism. With either mechanism, the junction will self-destruct if the current is not limited by external circuitry.

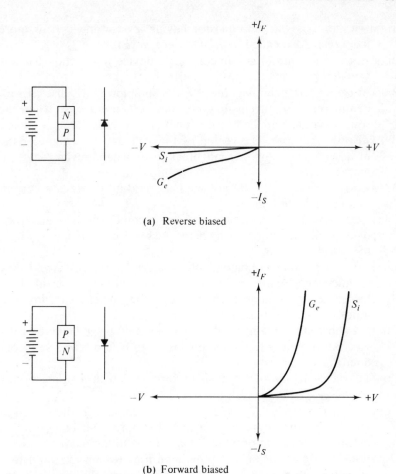

(a) Reverse biased

(b) Forward biased

Fig. 11–3   Forward and reverse biased PN junctions.

11.1 DEFINITIONS

The following terms are in general use.

ANODE: The positive electrode of a semiconductor. The electrode from which the forward current flows within the device.

BASE: A region of a semiconductor device into which minority carriers are injected.

BIPOLAR TRANSISTOR: A transistor having a base and two or more junctions. Also called *multijunction transistor*.

BLOCKING: The state of a semiconductor device or junction that essentially prevents the flow of current.

BREAKDOWN: The phenomenon of a PN junction that causes a sudden increase in current flow per unit voltage increase when the reverse voltage applied to the junction exceeds a critical level.

BREAKDOWN REGION: The region of the volt–ampere characteristic of a semiconductor beyond breakdown where reverse current increases rapidly.

BREAKDOWN VOLTAGE: The voltage measured at a specified current in a breakdown region.

CATHODE: The negative electrode of a semiconductor. The electrode of a semiconductor to which the forward current flows within the device.

CHANNEL: A region of semiconductor material in which current flow is influenced by a transverse electrical field, as in a field-effect transistor. Physically, a channel may be an inversion layer, a diffused layer, or bulk material.

CHIP: A substrate containing all the elements of a semiconductor device but requiring leads and enclosure to be a discrete semiconductor.

COLLECTOR: A region of a transistor through which a primary flow of charge carriers leaves the base.

COLLECTOR JUNCTION: A semiconductor junction normally biased in the reverse direction. The current through it can be controlled by the introduction of minority carriers into the base.

DOPING: Adding donor or acceptor impurities (called *dopants*) to a semiconductor material such as germanium or silicon to improve conductivity.

DRAIN: A region into which majority carriers flow from a channel of a field-effect transistor.

DUAL-GATE FIELD-EFFECT TRANSISTOR: A field-effect transistor having two independent gates, a source, and a drain.

ELECTRODE: Any external electrical connection to a region of a semiconductor device.

EMITTER: A region of a transistor from which charge carriers that are minority carriers in the base are injected into the base.

EMITTER JUNCTION: A semiconductor junction normally biased in the forward direction to inject minority carriers into the base.

FET (FIELD-EFFECT TRANSISTOR): A three-layer voltage-actuated semiconductor device. Conduction is by the flow of majority carriers through a conduction channel controlled by an electric field arising from a voltage applied between the gate and source terminals. (See MOSFET.)

FORWARD-BIAS: The bias which tends to produce current flow in the forward direction.

GATE: The control electrode of a field-effect transistor, comparable to the base of a bipolar transistor or the grid of a vacuum tube.

IGFET (INSULATED-GATE FIELD-EFFECT TRANSISTOR): A field-effect transistor having one or more gates that are electrically insulated from the channel. (Same as MOSFET.)

JUNCTION: The boundary between two regions in a semiconductor.

JFET (JUNCTION FIELD-EFFECT TRANSISTOR): A field-effect transistor that uses one or more gate regions that form a PN juntion(s) with the channel structure.

LED (LIGHT-EMITTING DIODE): A gallium-arsenide or gallium-phosphide semiconductor diode that emits light when energized by voltage.

MAJORITY CARRIERS: The predominant current carrier in a semiconductor; electrons in an N-type material and holes in a P-type material.

MOSFET (METAL-OXIDE-SEMICONDUCTOR FIELD-EFFECT TRANSISTOR): A field-effect transistor having one or more gates that are electrically insulated from the channel. Same as IGFET.

MINORITY CARRIERS: The less predominant current carrier in a semiconductor; electrons in P-type material and holes in an N-type material.

PELLET: A piece of a grown semiconductor crystal containing all of the active elements of a single semiconductor device but requiring leads and enclosure to be a discrete semiconductor device.

PUT (PROGRAMMABLE UNIJUNCTION TRANSISTOR): A P-N-P-N thyristor. With external resistors, it can generate a current–voltage characteristic similar to that of a unijunction transistor.

REFERENCE DIODE: A diode that is normally operated biased in the breakdown region of its voltage–current characteristic. A temperature-stable reference voltage of specified accuracy is developed across its terminals.

REVERSE BIAS: The bias that tends to produce current flow in the reverse direction.

SATURATION: A base-current and a collector-current condition resulting in a forward-biased collector junction.

SEMICONDUCTOR RECTIFIER DIODE: A semiconductor diode having an asymmetrical voltage–current characteristic that is used for rectification.

SEMICONDUCTOR SIGNAL DIODE: A semiconductor diode having an asymmetrical voltage-current characteristic that is used for signal detection.

SOURCE: A region from which majority carriers flow into the channel of a field-effect transistor.

SUBSTRATE: The supporting material on or in which the regions of a transistor are attached or made.

TUNNEL DIODE: A heavily doped junction diode with a low front-to-back resistance ratio and a negative-resistance region in the forward portion of the curve.

UJT (UNIJUNCTION TRANSISTOR): A three-terminal semiconductor device having one junction and a stable negative-resistance characteristic.

VOLTAGE–VARIABLE CAPACITOR DIODE: A silicon diode that utilizes the voltage–variable capacitance of a reverse-biased junction. The capacitance is inversely proportional to the applied voltage. See Chapter 2.

WAFER: A thin slice of semiconductor material in or on which a large quantity of semiconductor devices are formed. The wafer is afterwards diced (sliced, sawn, scribed, and broken) to obtain individual semiconductor devices.

11.2 BIPOLAR TRANSISTORS

A bipolar transistor (Fig. 11–4) consists essentially of a three-layer sandwich of N-type and P-type semiconducting materials. If the outer layers are P-type and the center is N-type, the transistor is a PNP; if the material types are reversed, it is an NPN. Terminals are attached to each of the three layers. The center layer is the base, one end is the emitter, and the other end is the collector. A bipolar transistor is a current-actuated device; its performance as an amplifier is expressed as the ratio of output current to input current, alpha or beta depending on how it is connected.

The two junctions in a bipolar transistor must be closely spaced in order to obtain useful electrical performance. The necessary non-uniform distribution of the adjacent P and N regions can be formed by *alloying, growing* or *diffusing* P and N impurities onto the surface or into the structure of the semiconductor pellet or wafer.

Alloy Transistors. An alloy transistor is made by alloying small amounts of donor or acceptor material into opposite sides of a thin doped semiconductor pellet to form emitter and collector regions, as shown in Fig. 11–5. The thickness of the pellet (the base),

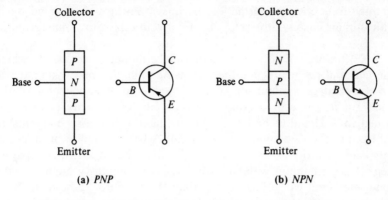

(a) *PNP*

(b) *NPN*

Fig. 11-4 Bipolar transistor diagrams and symbols.

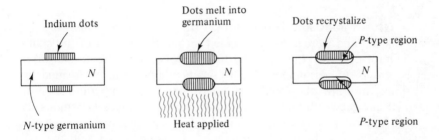

Fig. 11-5 Alloy transistor formation. *(Courtesy of General Electric Co.)*

the quantity of alloying material, the contact area of the alloying material, and the process temperature each affect the base width of the transistor, which determines most of the device characteristics.

The emitter and collector regions of an alloy transistor are very thin and have high electrical conductivity. This results in low saturation resistance, which is a measure of how good a short circuit the device becomes when turned on in a switching circuit.

A microalloy transistor achieves a thin base width by the pellet being selectively electrochemically etched to provide a thin spot or region. Metal is deposited on each side of the thin region and alloyed into the semiconductor to form the collector and emitter.

A variation, the microalloy diffused transistor, allows closer control of the transistor parameters. In this manufacturing process, a dopant is diffused into the semiconductor wafer before it is etched to form the thin regions. Both types of microalloy transistors have better high-frequency performance, because they are able to be made with thinner base widths than can a conventional alloy transformer.

Diffusion. Diffusion is a thermally induced process in which one material permeates another. In semiconductor processing, a wafer of doped germanium or silicon suitable for the collector of the transistor and one or a donor-acceptor pair of impurity elements is placed in a chamber and heated, causing the impurity elements to diffuse into the semiconductor material. A base region is produced if only one impurity is used; the emitter must be added by a separate step. If a germanium wafer is used with two impurity elements, the donor elements will diffuse faster than the acceptor, producing PNP structure; with a silicon wafer, the acceptor element will diffuse faster than the donor element, producing an NPN structure. The wafer is afterwards etched and sliced to yield multiple transistors. Diffusion is a slow process, but it allows close control of transistor configuration and electrical parameters.

Grown-Junction Transistors. In this type of transistor, the alternate P and N regions are formed during the growth of the semiconductor crystal rather than by being alloyed into the crystal after it is grown. In germanium, the P and N regions can be formed by varying the rate of growth by controlling the temperature; slow growth produces P regions, and rapid growth produces N regions. A single crystal can contain dozens of pairs of junctions. After the crystal is grown, the pairs of junctions are sawed apart, and the wafers are diced into several hundred pellets (Fig. 11–6). Each pellet needs only leads, mounting, and enclosure to become an NPN transistor. The process is not economically attractive for making PNP germanium transistors or for making silicon transistors.

Base width does not have any effect on the fragility of the device, unlike an alloyed transistor. Getting an adequate concentration of P-type impurities and the physical aspects of attaching a lead to the base limits the minimum width, which in turn restricts the frequency range.

Grown Diffused Transistors. In this manufacturing technique, impurity segregation during growth and diffusion are combined. The process can be used for both germanium PNP and silicon NPN transistors. In fabricating a silicon transistor, a lightly doped N-type crystal is dipped in molten silicon containing both N-type and P-type impurities, resulting in the growth of a highly conducting

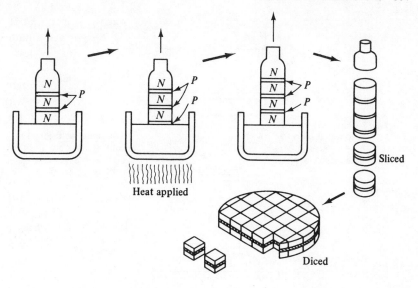

Heat applied

Sliced

Diced

Fig. 11–6 Grown-junction transistor formation.

N-type region containing a considerable quantity of P-type impurities, which at high temperature diffuse from the newly grown area into the lightly-doped N-type emitter region. This produces a P base region between a heavily doped N emitter and a lightly doped N collector. The technique provides good process control and a higher concentration of impurities in the base region than can be achieved by rate growing, resulting in improved high-frequency performance, limited only by the base lead attachment problem.

Mesa Transistors.   A mesa transistor has a small-area emitter and base that form a plateau or mesa on a larger collector, as shown in Fig. 11–7. The transistor is formed either by diffusing a base and emitter (double-diffused transistor) on the collector or by diffusing the base and alloying the collector onto the base (diffused-alloy transistor). The wafer is then etched to form a transistor structure and to expose a patch on the surface of the base for termination. The base can be as thin as desired, overcoming the frequency response limitations of alloyed and grown-junction transistors.

Mesa transistor collector resistivity is a compromise between the requirements for breakdown voltage (high resistivity) and low saturation voltage (low resistivity). Growing a high-resistivity epitaxial film on a low-resistivity pellet eliminates this need for compromise.

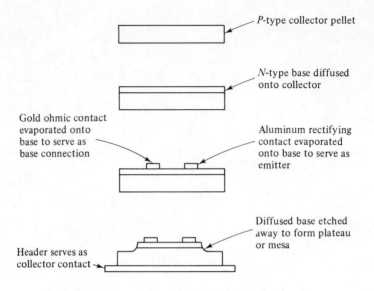

Fig. 11–7 Mesa transistor formation.

An *epitaxial film* is a film of single-crystal semiconductor material deposited on a single-crystal substrate. The atoms of the film are aligned with the substrate atoms to form a continuation of the substrate crystal. Epitaxial transistors have higher breakdown voltage, lower capacitance, and simultaneously lower saturation voltage than conventional mesas.

Silicon Planar Transistors.   In these transistors, both the emitter and the base are diffused (Fig. 11–8). First, silicon dioxide is formed on the silicon pellet to prevent the diffusion of impurities into the silicon. Then the silicon dioxide is etched through to form a mask for the diffusion of the base. The surface is again oxidized and etched for the emitter, which is then diffused. Aluminum is deposited on the base and emitter to form contacts. The oxide covers the junctions of the finished transistor, preventing gaseous contamination. This final oxide layer passivates the transistor, providing excellent electrical stability Epitaxial transistors can also be passivated in the same way.

11.3 FIELD-EFFECT TRANSISTORS

A field-effect transistor (FET) is a three-layer voltage-actuated device. The end layers are the source and drain; the center layer is the

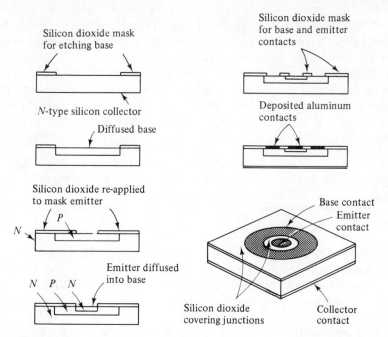

Fig. 11-8 Silicon planar transistor formation. *(Courtesy of General Electric Co.)*

gate. The electric field resulting from a control voltage on the gate modulates the flow of current through the semiconductor from source to drain. Its performance as an amplifier is expressed as *transconductance*, which is the ratio of output current to input voltage; this is the same term that is used for vacuum-tube gain. Some FETs have more than one gate. Symbols for FETs and MOSFETs are shown in Fig. 11-9.

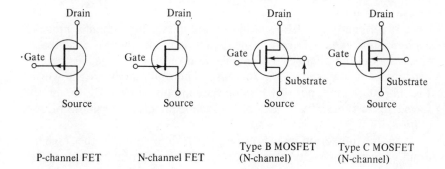

Fig. 11-9 FET and MOSFET symbols.

The field-effect transistor has several advantages over bipolar transistors. Noise is lower, input impedance is high (megohms), the FET is inherently more burn-out resistant, and it can be self-biased resulting in fewer circuit components. In fact, the FET behaves very much like a pentode vacuum tube, so much so that many circuits designed for small pentodes can use FETs with only minor modification. Table 11-1 compares characteristics of transistors and vacuum tubes.

TABLE 11-1   Comparison of Transistor and Vacuum Tube Characteristics

	Bipolar	JFET	MOSFET	Vacuum Tube
Warm-up Time	Negligible	Negligible	Negligible	Long
Aging Modes	No	No	Yes	Yes
Gate or Grid Current	High	0.1 nA	10 pA	1 nA
Input Impedance	Low	High	Very high	High
Bias-Voltage Temperature Stability	N/A	Good	Poor	Good
Overload Sensitivity	Good	Good	Poor	Very good
Noise	Low	Low	Variable	Low
Reliability	Good	Good	Good	Poor

There are two types of field-effect transistors; the Junction Field-Effect Transistor (JFET) and the Metal-Oxide-Semiconductor Field-Effect Transistor (MOSFET) also known as the Insulated-Gate Field-Effect Transistor (IGFET). Although the methods of manufacture of the two types are different, the principle of operation is the same—a "channel" current controlled by an electric field. However, device characteristics are different and require changes in external circuitry.

JFETs and MOSFETs both operate in two fundamental modes; *depletion* and *enhancement*. In the depletion mode, there is a decrease of carriers in the channel with application of gate voltage; in the enhancement mode, there is an increase. The differences are shown in Fig. 11-10. Some types of FETs operate in both modes. FETs are referred to as types "A," "B," or "C" depending on the mode of operation.

In the depletion mode (type A FET) operation, the highest drain current flows with zero gate voltage; the current is reduced by applying a reverse voltage to the gate. Type B FETs operate in a

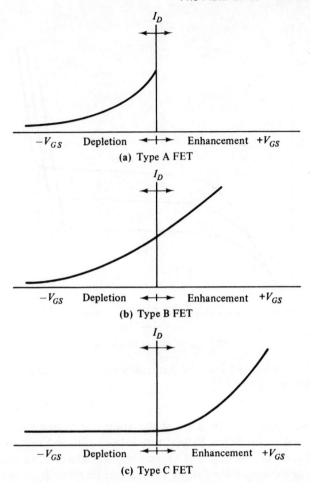

Fig. 11–10 FET operational modes. *(Courtesy of Motorola Semiconductor Group.)*

depletion/enhancement mode. Drain current can be increased by forward gate bias and reduced by reverse bias, providing a capability to handle large gate voltage swings. The type C FET operates only in the enhancement mode; drain current is extremely low with zero gate bias; drain current flows above some forward gate bias threshold level. Note the similarity of the $I_{DS}-V_{DS}$ curves in Fig. 11–11 to pentode vacuum tube plate-voltage/plate-current curves.

Junction Field-Effect Transistors. The JFET consists basically of a bar of doped silicon that behaves as a resistor, as shown in

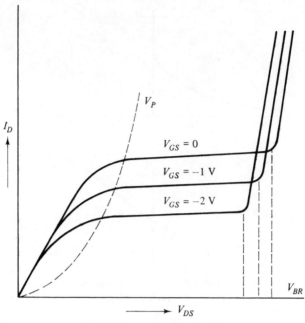

FET drain current characteristic

Fig. 11–11 Typical FET $I_{DS}-V_{DS}$ characteristic. *(Courtesy of Motorola Semiconductor Group.)*

Fig. 11–12. Current is injected at the source terminal; the other terminal is called the drain. P-type regions are diffused into an N-type doped silicon bar, leaving an N-type channel between the source and drain. (A complementary P-type device can be made by reversing the material types.)

A depletion region surrounds each of the P-N junctions when the junctions are reverse-biased. Increasing the reverse voltage spreads the depletion channels into the channels until they meet, creating an almost infinite resistance between the source and drain.

If instead, voltage is applied across the source and drain with zero gate voltage, the drain current in the channel sets up a reverse-bias along the surface of the gates parallel to the channel. Increasing the drain-source voltage spreads the depletion regions; when they meet, there is an effective increase in channel resistance that prevents any further increase in drain current. This voltage is the *pinch-off* voltage (V_p). Variations in drain current (I_d) with drain-source voltage (V_{DS}) for different gate voltages (V_{GS}) are shown in Fig. 11–11. Pinch-off occurs at lower levels as reverse-bias is increased.

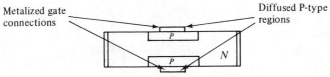

Fig. 11–12 FET Bar structure. *(Courtesy of Motorola Semiconductor Group.)*

In practical wafer processing, a single-side structure having a gate on one side only is used. The substrate is P-type material onto which an N-channel is grown epitaxially. A P-type gate is then diffused into the N-type epitaxial channel, with contact metalization completing the structure.

MOSFETs. Both the structure configuration and the control mechanism for a MOSFET are different from that of a JFET. Figure 11–13 shows the steps in the construction of a P-channel enhancement mode MOSFET. Two separate low-resistivity N-type regions (source and drain) are diffused into a high-resistivity P-type substrate. The surface is then covered with an oxide layer to insulate the channels from the gate. Over that a silicon nitride layer is applied to protect the oxide layer from sodium ion contamination, which would affect long-term stability. Holes are cut through the two layers to allow metal contacts to be attached to the source and drain; then a selective metal overlay is applied to the source and the drain and to the gate, which covers the entire channel area. The gate is insulated from the channel by the oxide layer. The drain and source are also isolated by the substrate. Without gate voltage, the drain-source substrate structure is comparable to two diodes back-to-back, which produces an extremely low drain-source current. A depletion mode MOSFET is made similarly, except that a moderate resistivity N-channel is diffused between the source and drain so that drain current flows with zero-gate bias.

MOSFETs are made in single-gate and dual-gate versions and

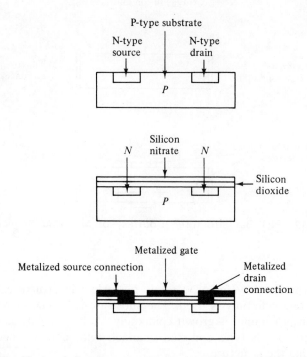

Fig. 11–13 N-channel enhancement-mode MOSFET formation. *(Courtesy of Motorola Semiconductor Group.)*

in a protected dual-gate version. A single-gate MOSFET provides extremely high input impedance and is capable of high power gains up to 250 MHz. Applications include RF amplifiers, mixers, and oscillators, audio and wide-band amplifiers, variable attenuators, choppers, and current limiters.

Dual-gate MOSFETs have two separate channels arranged in series. Each channel has an independent control gate, as shown in Fig. 11–14. Two gates provide advantages in mixer, product detector, remote AGC, demodulator, and modulator circuits as well as lower feedback capacitance and greater gain. Protected-gate MOSFETs have back-to-back diodes diffused into the substrate and electronically connected between each gate and the source. The diodes permit the device to handle wide signal swings and provide protection from in-circuit transients and against static discharge during handling.

Electrostatic discharges can occur when a device is picked up by its case; the handler's body capacitance is discharged to ground through the substrate-to-channel and channel-to-gate capacitances.

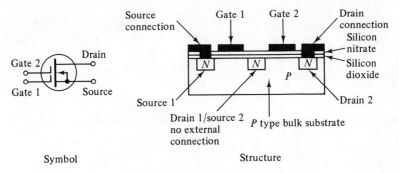

Fig. 11–14   Dual-gate MOSFET formation. *(Courtesy of RCA.)*

MOSFETs are susceptible to this kind of damage. Prior to assembly into a circuit, all leads should be kept shorted together by shorting springs or conductive foam. When removing the MOSFET from its carrier, the hand should be at ground potential. Soldering iron tips should be grounded; don't insert them into or remove them from circuits with the power on.

11.4 UNIJUNCTION TRANSISTORS

The unijunction transistor (UJT) is a three-terminal (emitter, base 1, base 2) semiconductor device. It has one junction and a stable negative resistance characteristic between the emitter and base 1 terminals when a positive bias voltage is applied between the base 1 and base 2 terminals. The symbol for the UJT and the static emitter characteristic curve are shown in Fig. 11–15.

The unique electrical characteristics of a unijunction include:

    1. A stable firing voltage, which is a fixed fraction of the interbase voltage.

    2. A very low value of firing current.

    3. A negative resistance characteristic that is stable with temperature and life.

    4. A high pulse-current capability.

There are several UJT structures. In the bar structure shown in Fig. 11–16, the mounting platform is a ceramic disc with a 10 mil slit in the center. A gold-antimony film is deposited on each side of the gap. A single-crystal N-type silicon bar is laid across the slit, and two ohmic base contacts are formed between the silicon and the

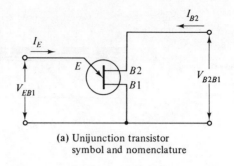

(a) Unijunction transistor
symbol and nomenclature

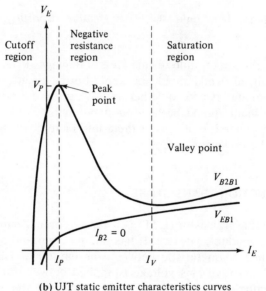

(b) UJT static emitter characteristics curves

Fig. 11-15 UJT symbol and static emitter characteristic curves. *(Courtesy of Motorola Semiconductor Group.)*

gold. A single P-type emitter is formed by alloying an aluminum wire onto the bar on the side opposite the base contacts, which is located off-center toward the base 2 contact so the device will not be symmetrical. The cube structure is similar.

In a different design more adaptable to automated production, the die is fabricated in a manner similar to that used for silicon planar passivated transistors (Fig. 11-17). A P-type boron emitter is diffused in an oxide-passivated die of high-resistivity N-type boron in a photoresist process. After the entire structure is again oxide-

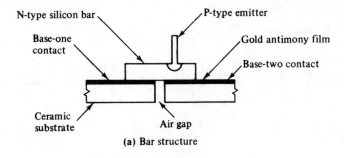

(a) Bar structure

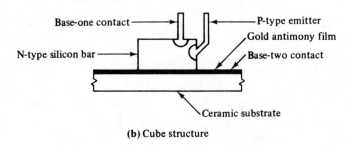

(b) Cube structure

Fig. 11–16 UJT bar and cube structures. *(Courtesy of Motorola Semi-conductor Group.)*

protected, windows are etched in the oxide, and the base 1 and the annular ring are formed by the diffusion of N-type phosphorus. The structure is again oxide-protected. The oxide is etched from the base 1 and emitter areas, aluminum is evaporated onto the die to make contact with the emitter and base 2, and gold is evaporated onto the bottom of the die to form the base 2 contact. A final etch shapes the contacts. In practice, these steps are performed on a wafer. After the final etch, the wafer is tested, scribed, and broken into several hundred dice.

As the emitter current (I_E) is increased, the voltage between the emitter and base 1 (V_{EB1}) increases to a peak point (V_p); with further increase in I_E, V_{EB1} decreases to the valley point (V_v). When the emitter current is increased into the saturation region beyond the valley point, the emitter current becomes a linear function of the emitter voltage.

Applications for the UJT include simple relaxation oscillators, trigger circuits, multivibrators, pulse generators and amplifiers, and frequency dividers.

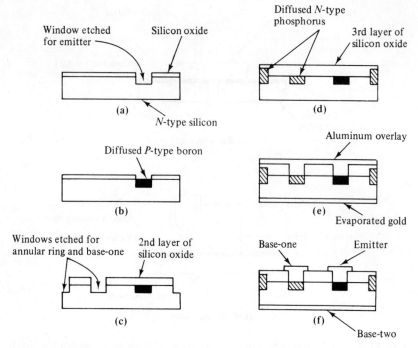

Fig. 11–17 UJT diffused construction. *(Courtesy of Motorola Semiconductor Group.)*

11.5 SEMICONDUCTOR RECTIFIER DIODES

A semiconductor rectifier diode is a two-terminal device that makes use of the rectifying property of a PN junction. Conduction is both asymmetrical and nonlinear. These diodes are made from many semiconductor materials, including silicon, germanium, selenium, and copper oxide. Characteristics are compared in Table 11–2. Diode symbols are shown in Fig. 11–18.

Almost all of the diodes used in electronic circuits today are silicon; the balance is mostly germanium with a few gallium arsenide. These diodes can be classed in a variety of ways: By base material (silicon, germanium); by structure (point contact, alloy, diffused, planar); by function (signal diode, switching diode, rectifier diode); and by current, voltage, or power dissipating rating. Typical power diode case styles are shown in Fig. 11–19.

A point-contact diode utilizes the rectifying contact between a sharp-pointed fine wire called a *catwhisker* and a pellet of semiconductor material called a *crystal*. Point-contact diodes have limited

TABLE 11-2   Rectifier Characteristics

	Germanium	*Silicon*	*Selenium*	*Copper-Oxide*
Peak Inverse Voltage				
Per Junction (max.)	200 v	2000 v	45 v	30 v
Reverse Current Leakage	Low	Very low	Moderate	High
Forward Voltage				
Drop Per Junction	0.7 v	0.9 v	1.0 v	0.4 v
Ability to Survive				
Voltage Transients	None	None	Excellent	Good
Operating Temperature (max.)	200°C	105°C	75°C	75°C
Life	Long	Long	8 to 12 years	Long

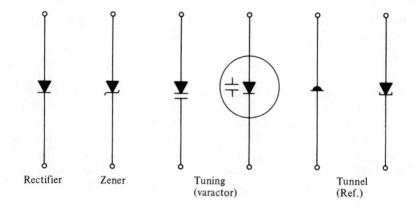

Rectifier   Zener   Tuning   Tunnel
                      (varactor)   (Ref.)

Fig. 11-18 Diode symbols.

power dissipation capability but have superior high-frequency characteristics due to low junction capacitance. Some types can operate at microwave frequencies.

Junction diodes are made by forming a PN junction in a pellet or wafer of doped germanium, silicon, or gallium arsenide. The material may be either N-type or P-type, and the junctions may be formed by either alloying, growing, or diffusion. Junction diodes are made in a variety of sizes and ratings. They are more rugged than equivalent point-contact diodes, but their maximum operation frequency is much lower principally because of high junction capacitance.

Planar diodes are junction diodes made by diffusing either an

Fig. 11–19 Typical power diode case styles. **(a)** DO-ı, **(b)** DO-15, **(c)** TO-3, **(d)** TO-66, **(e)** DO-5. *(Courtesy of RCA.)*

N-type impurity into a P-type wafer or a P-type impurity into an N-type wafer, as shown in Fig. 11–20. Connections to both anode and cathode are made on the same face of the wafer.

A planar passivated diode is a type of planar diode that has had a silicon dioxide film formed on the surface to protect the junction materials from contamination or short circuit. Holes are etched in the oxide layer for connections to the anode and cathode.

11.6 ZENER DIODES

These devices are also known as *avalanche diodes* and *breakdown diodes*. A zener diode is a two-layer semiconductor device that has a sudden, very abrupt rise in current flow when an applied reverse-bias voltage is increased to the region of the breakdown voltage. With further increase in the reverse-bias voltage, the voltage drop across the device junction remains essentially constant (within the heat-dissipation rating of the device). When forward-biased, a zener diode behaves as an ordinary rectifier (Fig. 11–21). Zener diodes are used for voltage regulation and overvoltage protection and to provide a reference voltage.

The breakdown voltages for zener diodes range from 1.8 volts to 200 volts with power ratings from 1/4 watt to 50 watts. True

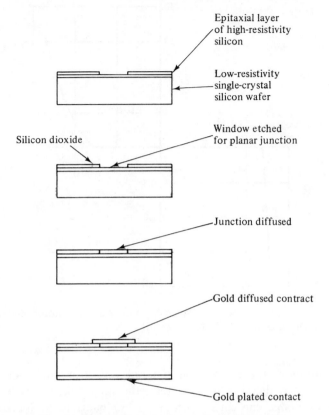

Fig. 11–20 Epitaxial planar diode formation.

zener breakdown predominates for breakdown voltages only up to five volts; above eight volts, the avalanche mechanism predominates; in between, both occur.

There are two kinds of zener diodes: Regulator diodes and reference diodes. Regulator diodes are used in applications such as power supplies where a nearly constant dc output voltage is desired despite changes in input voltage and load current. Regulator diodes are temperature sensitive. Reference diodes are not temperature sensitive and provide tighter control.

A forward-biased PN junction has a negative temperature coefficient; a reverse-biased junction has a positive temperature coefficient. Judicious combination of forward- and reverse-biased junctions in the same device produces a reference zener diode with a very low temperature coefficient.

Zener diodes are made by several processes. P-type material

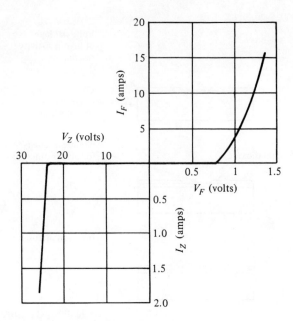

Fig. 11–21 Typical zener diode voltage-current characteristic. *(Courtesy of Motorola Semiconductor Group.)*

can be diffused into an N-type material wafer, or the reverse, and the wafer then diced. Unfortunately, the high field strength across the junction edge can cause surface contaminants to be ionized and swept into the junction area producing erratic performance.

In a planar process, the junctions are on the surface of the chip and are protected by a passivating covering of silicon oxide; however, some ionization still can occur because of the high electric field strengths. Another approach is to place a high-resistance epitaxial N-type layer on an N-type chip, producing a lower field in the covering oxide passivating layer; in addition, the surface of the oxide layer is metalized to reduce ion migration.

Alloy-diffused zener diodes are also made. A boron P-type region is diffused into an N-type wafer. The surface is passivated and then etched to allow aluminum to be alloyed into the base through the diffused junction.

11.7 THYRISTORS

A thyristor is any bistable semiconductor switching device having three or more PN junctions that can be switched from a high-impedance OFF state to a low impedance ON state by an external signal or voltage. Thyristor performance is comparable to that of a thyratron tube. There are several types of thyristors: Silicon controlled rectifiers (SCRs), silicon controlled switches (SCSs), and triacs are the principal varieties. Typical thyristor case styles are shown in Fig. 11–22.

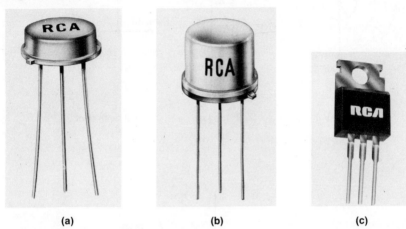

(a)   (b)   (c)

Fig. 11–22 Typical thyristor case styles. **(a)** TO-5, **(b)** modified TO-5, **(c)** TO-220AB. *(Courtesy of RCA)*

Silicon Controlled Rectifier (SCR). The SCR is a four-layer P-N-P-N bistable switch with three terminals: Anode, cathode, and gate, as shown in Fig. 11–23. It functions as a diode in the reverse direction and as an electronic switch in the forward direction. With no control signal on the gate, current flow between the anode and cathode is blocked in both directions. A positive control pulse on the gate triggers the SCR into conduction in the forward direction.

The SCR operates as a latched switch. Once the SCR has been triggered into conduction, the gate loses all control. The SCR can only be turned OFF by interruption, removal, reduction, or reversal of the anode voltage.

Methods of construction vary, dependent on the power-handling capability for the particular device and its intended application, as shown in Fig. 11–24. All begin with P-type impurities being diffused into oxide-masked N-type silicon wafers. For small, low current SCRs, the wafer can be converted to a P-N-P-N wafer with a second diffusion following oxidation and masking. The wafers are then diced into usable SCR pellets.

High-current SCRs requiring larger pellets are usually diced af-

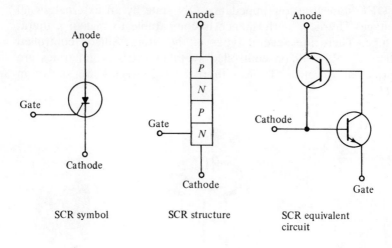

SCR symbol   SCR structure   SCR equivalent circuit

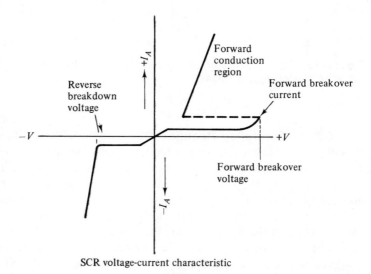

SCR voltage-current characteristic

Fig. 11-23 SCR symbol, structure, equivalent circuit and voltage-current characteristic.

ter the first diffusion. The final N-type layer is formed by alloying a gold-antimony preform into the peliet.

SCRs designed for consumer and light industrial applications are soft-soldered to the case bottom or to a copper stud for larger

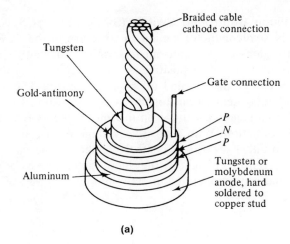

(a)

Alloy-diffused SCR pellet

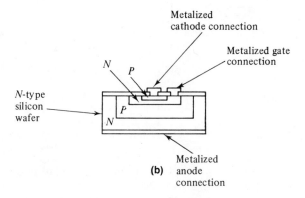

(b)

Fig. 11–24   SCR construction. **(a)** Alloy-diffused SCR pellet structure used in high-power devices, **(b)** planar passivated SCR structure used in lower power devices. *(Courtesy of General Electric Co.)*

models. Large SCRs, and any SCRs for applications where temperature extremes or severe thermal cycling is anticipated, have the P-N-P-N pellet first brazed between plates of molybdenum or tungsten; then the bottom plate is hard-soldered to a copper stud. Small lead-mounted types have only the bottom plate, which then becomes part of the SCR case. For some low-current SCRs, the silicon oxide layer is depended on for sealing, and the SCR is encapsulated with leads and heat-sink in a plastic molded case.

Silicon Controlled Switch (SCS). The SCS is a four-layer P-N-P-N semiconductor similar in structure to the SCR except that all four regions are brought out to terminals. It can be triggered into conduction by either a positive or negative pulse. Unlike the SCR, the SCS can be also turned off by gate control. The symbol, structure and equivalent circuit for the SCS is shown in Fig. 11–25.

With a negative voltage on the anode, the SCS will not conduct in either direction. With a positive voltage on the anode, it will conduct with either a positive control voltage on the cathode gate or a negative control voltage on the anode gate. The SCS is a versatile device and can be operated in many modes.

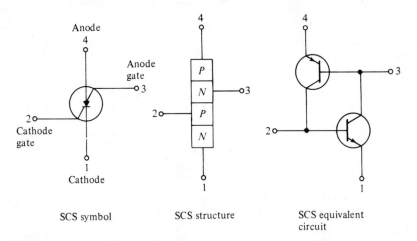

| SCS symbol | SCS structure | SCS equivalent circuit |

Fig. 11–25 SCS symbol, pellet structure and equivalent circuits. *(Courtesy of General Electric Co.)*

Triac. A triac is a three-terminal, five-layer N-P-N-P-N semiconductor switch having an ON–OFF switching characteristic similar to that of an SCR except that the triac will conduct current in either direction (Fig. 11–26). The gate signal may be either positive or negative, and the triac can switch both ac and dc. The triac may be thought of as a pair of complementary SCRs in parallel.

Diac. A diac is not a thyristor. A diac is a two-terminal, three-layer silicon switch that operates in avalanche breakdown once an applied ac or dc voltage reaches a critical value. Below that value, only a small leakage current flows. Diacs are used in simple control circuits and to trigger thyristors (Fig. 11–27).

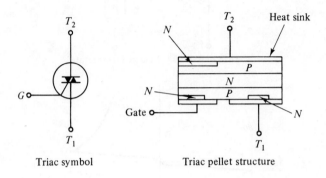

Triac symbol   Triac pellet structure

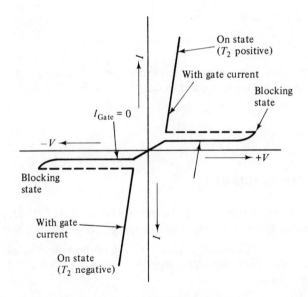

Triac voltage – current characteristic

Fig. 11–26 Triac symbol, pellet structure and voltage-current characteristic. *(Courtesy of General Electric Co.)*

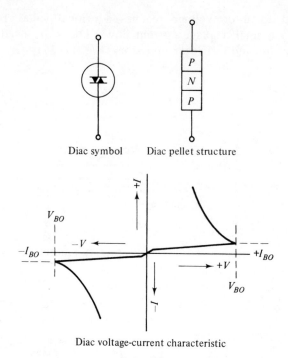

Diac symbol   Diac pellet structure

Diac voltage-current characteristic

Fig. 11–27 Diac symbol, pellet structure and voltage-current characteristic. *(Courtesy of General Electric Co.)*

SUGGESTED READINGS

Galloway, J. H., *Using the Triac for Control of AC Power,* 200.35. General Electric Company, Syracuse, N. Y., July 1970.

General Electric Company, Electronic Components Sales Department, *Silicon Controlled Rectifier Manual,* 3rd ed. Syracuse, N. Y., 1964.

General Electric Company, Semiconductor Products Department, *Transistor Manual,* 5th ed. Syracuse, N. Y., 1960.

General Electric Company, Semiconductor Products Department, *Transistor Manual,* 6th ed. Syracuse, N. Y., 1962.

General Electric Company, Semiconductor Products Department, *Transistor Manual,* 7th ed. Syracuse, N. Y., 1964.

Graham, D. R., *Using Low Current SCR's*, 200.19. Semiconductor Products Department, General Electric Company, Syracuse, N. Y., January 1967.

Lenk, John D., *Handbook of Electronic Components and Circuits.* Prentice-Hall, Inc., Englewood Cliffs, N. J., 1974.

Lowry, H. R., J. Giorgis, E. Gottlieb, and R. C. Wischedel, *Tunnel Diode Manual,* 1st ed. Semiconductor Products Department, General Electric Company, Liverpool, N. Y., 1961.

Motorola Semiconductor Products, Inc., Applications Engineering, *Circuit Applications for the Triac,* AN–466. Phoenix, Ariz., 1971.

Motorola Semiconductor Products, Inc., Applications Engineering, *Field Effect Transistors in Theory ,and Practice,* AN–211A. Phoenix, Ariz., 1973.

Motorola Semiconductor Products, Inc., Applications Engineering, *Theory and Characteristics of the Unijunction Transistor,* AN–293. Phoenix, Ariz., 1972.

Motorola Semicondutor Products, Inc., Applications Engineering, *Silicon Rectifier Handbook.* Phoenix, Ariz., 1973.

Motorola Semiconductor Products, Inc., Applications Engineering, *Zener Diode Handbook.* Phoenix, Ariz., 1967.

RCA Corp., *MOSFET Product Guide.* Somerville, N. J., October 1970.

RCA Corp., *Thyristors/Rectifiers,* SSD–206C, 1975 Databook Series. Somerville, N. J., 1974.

Stasior, R. A., *Silicon Controlled Switches,* 90.16. General Electric Company, Syracuse, N. Y., June 1964.

Sylvan, T. P., *The Unijunction Transistor Characteristics and Applications,* Application Note, 90.10. Semiconductor Products Department, General Electric, Syracuse, N. Y., May 1965.

Texas Instruments Incorporated, Components Group, *The Transistor and Diode Data Book for Design Engineers,* 1st ed. Dallas, Tex., 1974.

Walters, C. Kent, and Ronald N. Racino, *Design Considerations and Performance of Motorola Temperature-Compensated Zener (Reference) Diodes,* AN–437B. Motorola Semiconductor Products, Inc., Phoenix, Ariz., 1974.

Index